INTELLIGENT *Design*
or
NON-INTELLIGENT *Design?*

Searching for Answers

KEITH ERICSON

Illustrations by **COURTNEY COTTON**

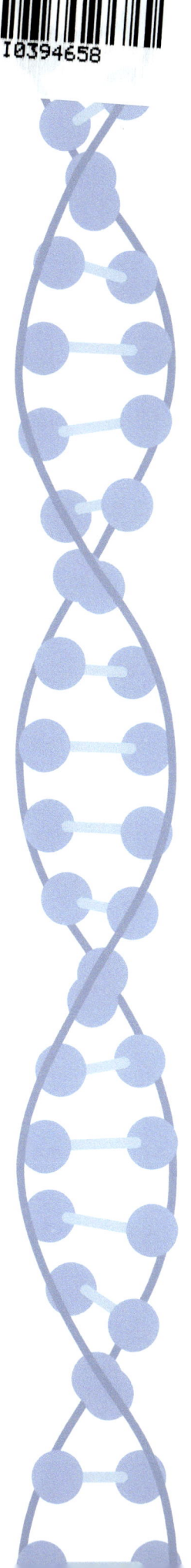

Copyright © 2013 Keith Ericson.

All rights reserved. No part of this book may be used or reproduced by any means, graphic, electronic, or mechanical, including photocopying, recording, taping or by any information storage retrieval system without the written permission of the publisher except in the case of brief quotations embodied in critical articles and reviews.

All images © Courtney Cotton.

LifeRich Publishing books may be ordered through booksellers or by contacting:

LifeRich Publishing
1663 Liberty Drive
Bloomington, IN 47403
www.liferichpublishing.com
1 (888) 238-8637

Because of the dynamic nature of the Internet, any web addresses or links contained in this book may have changed since publication and may no longer be valid. The views expressed in this work are solely those of the author and do not necessarily reflect the views of the publisher, and the publisher hereby disclaims any responsibility for them.

ISBN: 978-1-4897-0066-7 (sc)
ISBN: 978-1-4897-0068-1 (hc)
ISBN: 978-1-4897-0067-4 (e)

Printed in the United States of America.

LifeRich Publishing rev. date: 12/17/2013

PREFACE

This is not a book about religion. Instead, it is a book about science and logic and common sense. We deny in this book that humans are the most intelligent beings in this vast universe. This denial is based on the fact that there is too much evidence of design in so much that we see everyday. The question we pose is "Is there intelligence behind this design, or is everything the result of uncontrolled accidents happening over and over through millions of years to produce, finally, the extremely complicated human body and other organisms inhabiting this earth today?"

In this work we steer away completely from using the various terms that name or describe supreme beings as authors of major religions. We use the term "powerful intelligence" to refer to the thinking and planning behind all of the designed beings/creatures on this planet.

In this book you will see the term "pure evolutionist" used many times. This term describes the person who believes that there is no intelligence behind the design of the human body - that everything happened by chance or by accident. The pure evolutionist probably believes that humans are the most intelligent beings in this universe.

Also, we will not diagnose any illness nor will we offer solutions to problems. All of the investigations we perform as we look at the human body are done with the assumption that we are looking at a healthy body or body part.

This book is not one that the medical student could use as a text to help him/her through the rigorous study necessary to become a doctor. Instead, it is intended as a basic source of information to answer the question "can a complicated body part design itself, can the rest of the body design a part that it needs, or is there intelligence behind the design of the human body?"

The term "need" appears many times when an evolutionist explains how different body parts came into existence. For instance, if the body needed a way to bring oxygen into its self in order to remain alive, the evolutionist will tell us that that body somehow recognized the need and through millions of years, that "oxygen-intake system" developed. And no matter how intricately that system is designed, it happened on its own without any outside help; it just developed on its own because it was needed. We question this kind of explanation throughout this book.

Finally, we advance the idea that there cannot be such a concept as non-intelligent design. Anything that shows design also cries out that there is some kind of intelligence behind that design.

So we encourage our readers to examine carefully the science the medical profession deals with every day as this profession works with the human body. And as we examine this science, we encourage our readers to use logic and common sense in answering the question "can there be such a thing as 'non-intelligent' design"?

INTRODUCTION
Why Consider Intelligent Design?

Before we begin looking at some of the controversy surrounding design versus chance or accident we must acknowledge the fact that many honest, capable, intelligent people accept the idea that no intelligence of any kind played a role in making the human race what it is today. But when pure evolution is presented as indisputable fact and rules out any possibility of an intelligence of some kind designing our amazing, complicated bodies, it's natural that another segment of the population will take issue with that presentation. Although we poke a little fun at the idea that we are here by chance or accident, we do not intend to insult any who accept the doctrine of pure evolution.

Theory tells us that millions and millions of years ago the earth was a big red-hot ball. Of course there was no life of any kind - it was just too hot. But as the earth cooled some warm liquid appeared with all kinds of minerals and dirt on and under the surface. Some time later – approximately a few million years - some kind of a miracle happened and a couple microscopic minerals got together and a living cell appeared. This event was repeated over and over - lots of living cells - because that liquid (some call it primordial soup) was cool enough to allow these one-celled "things" to exist. So theory tells us that they found some way to take nourishment and survive and multiply. And all of this supposedly happened without any guidance or planning or intelligence behind any of it.

Then, more miracles - these one-celled things began uniting and growing and before you know it (maybe a couple more million years?) we've got some blobs of multi-celled creatures! Also, some of these multi-celled things became plants instead of animals. So while that primordial soup is still the main liquid, some of these multi-celled animals began climbing out of the soup - maybe they rolled out because they didn't have appendages yet. But we've got a slight problem with this picture. Those that stayed in the soup needed gills in order to absorb oxygen, so they developed gills as soon as they became multi-celled things. But the ones that rolled out onto dry land needed lungs in order to survive, so immediately they got their lungs!

But, what are they going to do with their oxygen? They have no blood stream to transport it, they have no heart, no brain - they are just blobs with no purpose yet for their existence, and no reason for their existence. And when we look at how unlikely it is that they will live very long, we will probably have to wait for another round of uncontrolled miracles in that primordial soup.

We could go on and on with these ideas, but it gets more and more absurd as we try to explain our existence as coming from some kind of primordial soup. Because as we look at so many needs that had to be met, we have to wonder if it is remotely possible that we are what we are today because non-living matter became living matter by accident after accident with no intelligence behind it in any way.

Common sense tells us to take an in-depth look at what we are today and come up with a rational explanation instead of the picture presented above.

CONTENTS

1. Heart 1
2. Blood, Arteries and Veins 6
3. The Nervous System - Brain and Spinal Cord 10
4. Lungs 16
5. Digestive System 20
6. Kidneys 25
7. Sense of Hearing 29
8. Senses of Taste and Smell 32
9. Sense of Vision 35
10. Skeleton, Bones and Teeth 39
11. Muscles 44
12. Immunity and Lymphatics 47
13. Smoking 51
14. Cells, Chromosomes, DNA and Genes 53
15. What Motivates the Medical Student? 56

Summary Statements 58

Test on Intelligent Design 62

Answers to Test 65

Chapter 1
HEART

The heart, the organ that pumps blood to the entire body, lies in the chest behind the sternum (breast bone). It is made up mostly of muscle that beats about 70 times each minute if not under any stress such as exercise.

The heart has four chambers, the right atrium and right ventricle, and the left atrium and left ventricle (figure 1).

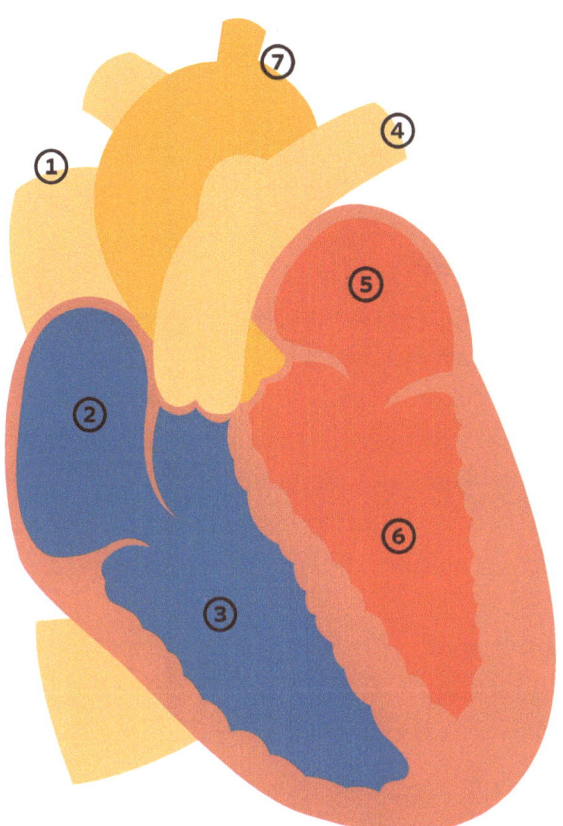

Chapter 1, Figure 1

1. Superior vena cava - brings deoxygenated blood from upper body to right atrium. The inferior vena cava (not shown) brings blood from the lower body also to the right atrium.
2. Right atrium - Deoxygenated blood enters this chamber from all parts of the body.
3. Right ventricle - The deoxygenated blood enters this chamber from the right atrium and goes to the lungs to receive oxygen and give up carbon dioxide.
4. Pulmonary artery - Takes deoxygenated blood to the lungs.
5. Left atrium - Oxygenated blood enters this chamber from the lungs and goes to the left ventricle.
6. Left ventricle - Blood is pumped from this chamber through the aorta to all parts of the body.
7. Aorta

The right atrium receives blood through two veins that bring blood from the entire lower and upper body. The blood is then pumped to the right ventricle from which it is pumped to both lungs where it is relieved of carbon dioxide and receives oxygen (the blood going to the lungs is called deoxygenated blood and the blood coming from the lungs is called oxygenated blood). The blood that comes from the lungs goes to the left atrium, is pumped to the left ventricle and from there it is pumped to the main artery of the body, the aorta, which moves blood to the entire body. This is a continuous action that never stops as long as our body is healthy.

We do not control our heartbeat. If we are exercising and require more oxygen, our brain tells our heart to beat faster to send a greater supply of deoxygenated blood to the lungs, where it will be replenished with oxygen, go back to the heart and then through the arteries to the rest of the body.

Since we don't control our heartbeat, let's look at how the heart continues to beat on its own. (This gets a little technical, but it shows how complicated the process is.) In the right atrium is a small cluster of cells called the sino-atrial node (the SA) (figure 2). This little cell cluster is known as the heart's pacemaker. Electrical impulses from the SA send messages to other nerves in the heart and all of the four chambers beat on a regular schedule if the heart is working properly. The right atrium and the left atrium beat at the same time and both ventricles beat a fraction of a second later. This happens because the Sino-atrial node sends a signal to the atrioventricular node (AV) (figure 2), which then sends signals to the ventricals telling them to beat. They actually beat about one tenth of a second after each atrium. A doctor listening to a stethoscope hears these two beats, but if we check our own pulse we feel only one beat because we feel the beat of the left ventricle as it moves blood into the aorta. When we feel our pulse we are feeling an artery. Blood moving through veins moves steadily and does not show a beat.

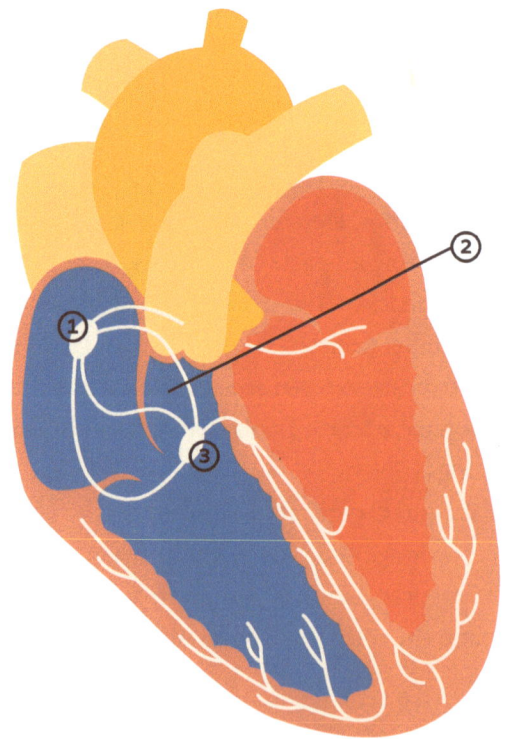

Chapter 1, Figure 2

Cut away view of the heart showing the sinoatrial node (SA) and atrioventricular node (AV).
The sinoatrial node sends signals to each atrium to beat and also sends signals to the atrioventricular node to beat. The ventricals beat about one tenth of a second later than each atrium. When we check our pulse at our wrist, we only feel the beat of the left ventrical as it pumps blood to the aorta.
1. Sinoatri node
2. Right atrium
3. Atrioventricular node

We could become just slightly misinformed at this point. If we think the heart never rests we are wrong. The heart muscle rests just a fraction of a second between beats. Check your pulse - there is a slight pause between beats, which indicates the heart has rested between those beats.

The brain enters the picture when the body is stressed and needs more oxygen and "tells" the heart to beat faster.

Something else happens that actually keeps the heart alive. The heart has a system of arteries called coronary arteries, which supply blood to the muscle and covering of the heart itself (figure 3). At the point where the aorta leaves the left ventricle, three coronary arteries are connected to the aorta and supply oxygenated blood to the heart itself. Veins then transport this "used" blood to the right atrium. From there the blood continues its normal journey finally to the lungs where it gives up carbon dioxide and receives oxygen.

Chapter 1, Figure 3

This illustration shows a simplified drawing of the blood supply to the heart itself. Where the aorta first comes out of the heart, three smaller arteries branch off and supply the heart muscle with oxygenated blood. This drawing only shows the three arteries as they branch off from the aorta. They actually form many smaller branches and cover the entire heart. Veins are also in place to receive blood containing carbon dioxide. This deoxygenated blood eventually reaches the right atrium, as does all deoxygenated blood.

1. Aorta
2. Main veins bringing deoxygenated blood from the entire body
3. Right coronary artery
4. Front coronary artery
5. Left coronary artery - This artery branches off and sends blood to the back of the heart. It works the same as the right and front coronary arteries.

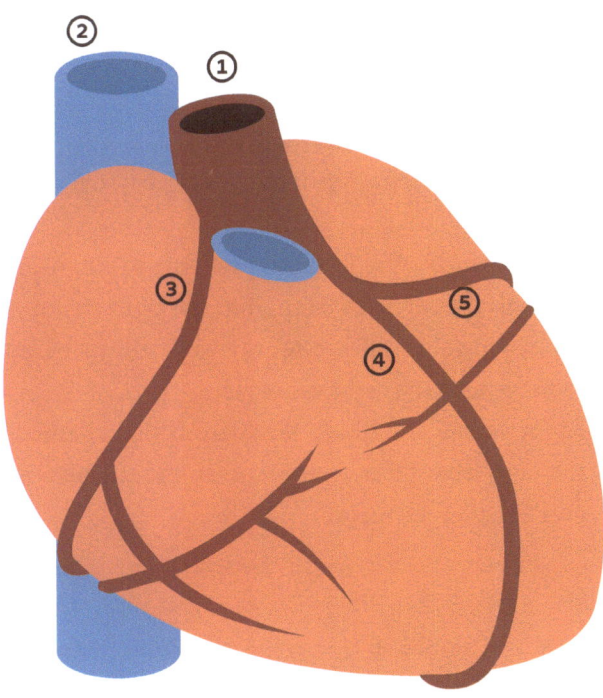

There are four valves in the heart that must work properly if the heart is to function as it should (figure 4). The valve between the right atrium and the right ventricle is called the tricuspid valve. It opens to let blood flow to the ventricle, and closes when the ventricle pumps blood to the lungs. The artery taking blood from the right ventricle to the lungs is called the pulmonary artery and the valve separating the ventricle from the artery is called the pulmonary valve. This valve opens when the ventricle pumps and closes when the pumping action stops. Both of these valves close at the right time to prevent back flow of blood. In the left side of the heart the mitral valve separates the atrium from the ventricle and the aortic valve separates the ventricle from the aorta. These two valves work the same as the valves in the right side of the heart. They open to let blood flow through and close to prevent back flow.

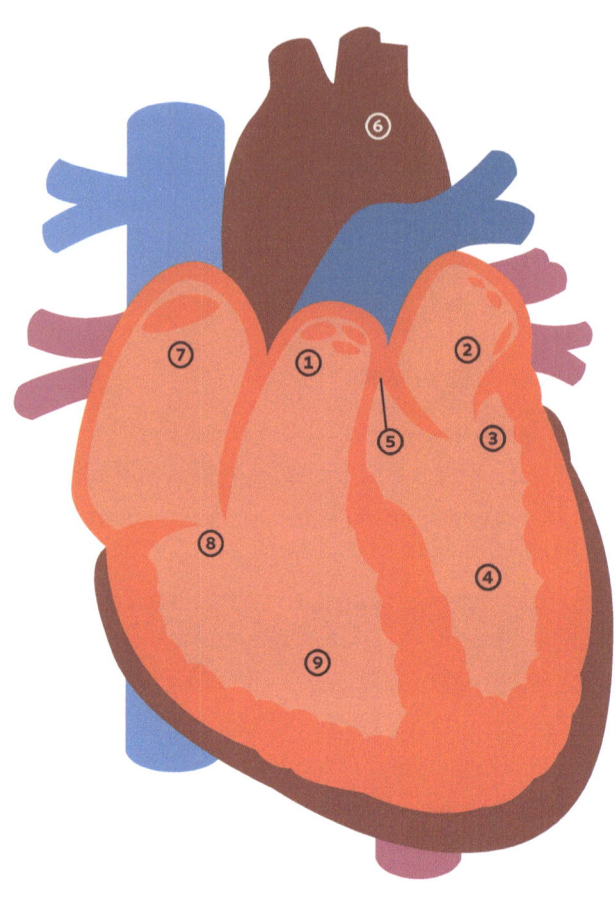

Chapter 1, Figure 4

Valves of the heart - This drawing shows a cut away view of the heart. You will see the aorta, the four chambers of the heart (the pumps), and the four one-way valves.

1. Pulmonary valve - deoxygenated blood flows through this valve, through the pulmonary arteries to both lungs where it will pick up oxygen and get rid of carbon dioxide.
2. Left atrium - blood comes to this chamber from the lungs.
3. Mitral valve - Blood flows through this valve into the left ventrical.
4. Left ventrical
5. Aortic valve - Blood flows through this valve into the aorta, which moves it to the entire body.
6. Aorta
7. Right atrium - Blood finally returns to the heart through the venous system into the right atrium.
8. Tricuspid valve - This one-way valve allows blood to move into the right ventricle.
9. Right ventricle - Blood flows from this chamber through the pulmonary valve and begins its journey all over again.

Let's review, starting with the right ventricle.

The right ventricle sends deoxygenated blood to the lungs where it gives up carbon dioxide and receives fresh oxygen (it is now oxygenated blood). This fresh blood enters the heart through the left atrium, is pumped to the left ventricle, and leaves the heart through the aorta and the blood flows to all parts of the body. It returns to the right atrium of the heart through veins as deoxygenated blood, is pumped to the right ventricle and the "tour" starts all over again. During this process the heart keeps beating because, in the right atrium, is the cell cluster, the SA, which sends nerve signals to each atrium and to the AV which sends signals to the ventricles. These four pumps (chambers) continue beating as they are designed to do. When the body is stressed, the brain sends a signal to the heart to beat faster because oxygen is being used up at a faster rate. And while all of this is happening, coronary arteries receive blood from the aorta and send that blood to the entire surface of the heart. Then veins take the deoxygenated blood back to the right atrium. Along with all of this, the heart has four valves that open and close at needed times as the four chambers (pumps) move blood on its journey. And these valves are designed to prevent the back flow of blood between each atrium and ventricle, and between the ventricles and the pulmonary artery and the aorta.

Obviously, the lungs play an important part in the operation of the heart. Without the lungs, the heart would be useless. We'll take a close look at the lungs in chapter four.

As we examine the design of the heart, its structure, its stamina and its role in keeping the human body functioning, it becomes apparent that it is a complicated combination of muscle, arteries, veins and valves that keep us alive and well.

Chapter 2
BLOOD, ARTERIES AND VEINS

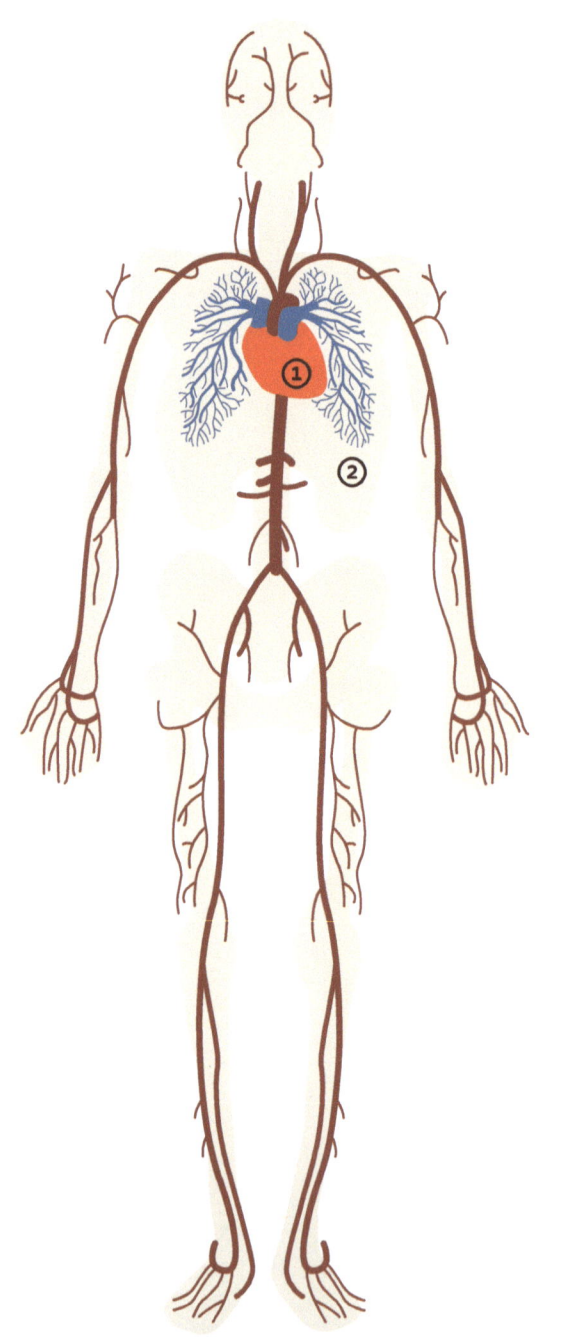

It makes sense to discuss the blood that moves to all parts of the body at the same time we discuss the network of arteries and veins through which the blood moves. We know from chapter one that blood leaves the heart from the left ventricle through the aorta at the upper end of the heart. From here the blood travels to all parts of the body. We also know that this blood is oxygenated blood because it has just come from the lungs where it gave up carbon dioxide and picked up oxygen. (How this happens is discussed in chapter four which covers the lungs.)

Arteries go to all parts of the body full of oxygenated blood and veins come from all parts of the body with deoxygenated blood. Figure one is a diagram of the network of arteries carrying oxygenated blood. A drawing of the veins (figure 2) is almost the same although veins seem to vary a little in their positioning from body to body while arteries show almost the same pattern or layout in all people.

Chapter 2, Figure 1

In this drawing we show the main arteries in the human body. These arteries divide into smaller and smaller arteries until they connect with capillaries. At this point they are called arterioles. This drawing shows only the main arteries.
1. Heart
2. Lungs showing deoxygenated blood

arterioles. Another kind of design appears at this point. The arterioles connect with capillaries, which in turn connect with the system of veins, which will take the blood back to the heart (figure 3). However these veins are so small that they are called venules, but as they connect with other venules they become larger and then are known as veins.

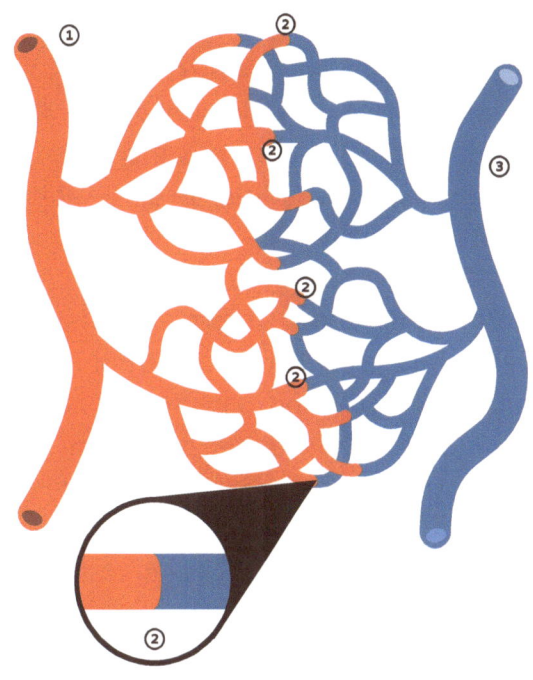

Chapter 2, Figure 3

This illustration shows arteries and veins with microscopic capillaries connecting the blood vessels. The capillaries are located where the arterioles and venules meet. It is in the capillaries that oxygenated blood is exchanged for deoxygenated blood. The oxygen-rich blood is absorbed into the body's cells because of the work of the capillaries and the capillaries also pull carbon dioxide from the cells and "deposit" it in the venules. This "used blood" travels through the veins back to the heart. The oxygenated blood is shown in red and the deoxygenated blood is blue.
1. Arteries carry oxygenated blood.
2. Capillaries are located where the arterioles and venules meet.
3. Veins transport carbon dioxide-rich blood back to the heart.

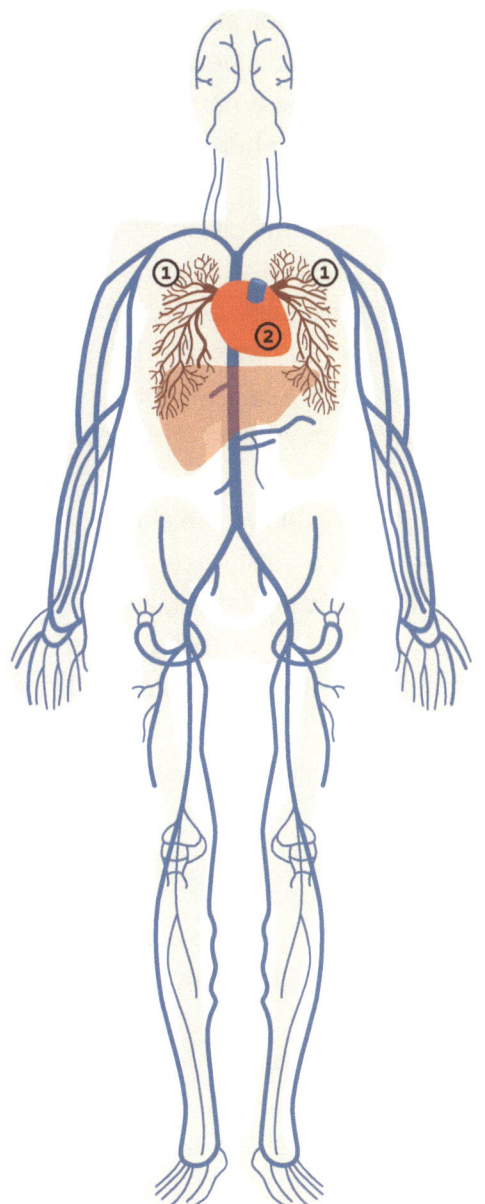

Chapter 2, Figure 2

This drawing shows only the main veins in the human body. Tiny veins, called venules, bring deoxygenated blood from capillaries to the main veins and finally back to the heart.
1. Lungs showing oxygenated blood
2. Heart

The arteries divide as their distance from the heart/aorta increases. They become smaller and smaller until they reach the tissue they will supply with oxygen. The smallest of the arteries are called

But let's go back to the capillaries - because it is here that the blood deposits oxygen and picks up carbon dioxide. All through the body this process is constantly taking place. If you take your pulse at your wrist, every beat you feel means that capillaries are exchanging oxygenated blood with deoxygenated blood. When you are sleeping your heart beats about 70 - 80 times per minute and this exchange is taking place constantly.

So the deoxygenated blood travels back to the heart and finally enters the right atrium, then to the right ventricle, to the lungs, then back to the left atrium, the left ventricle, up to the aorta and it is on its way again.

So, all of this seems simple enough - we've "traveled" with a drop of blood from the heart, through our body, and back to the heart. But many different things happen in different parts of our body while the blood is moving. For instance, we remember that we feel no pulse in our veins, which means we feel no pumping action. Blood does flow through our veins but it is helped on its way by the muscles in our body, as well as the pumping action of the heart. The longest veins are in our legs. When we are standing and blood is moving toward our heart, gravity is pulling against that flow. But there is another important design feature that "fights" gravity. The muscles contract against the veins every time we move our legs forcing blood to move. And to keep it from moving in the wrong direction there are one-way-valves, designed into our veins so the blood can move only in one direction - toward the heart.

Muscular action all through our body helps move blood but our legs give us the most striking example.

Blood aids in the digestive process in the stomach and small intestines and the liver but this action is covered in other chapters.

We've taken a basic look at the circulatory system. Obviously we haven't covered all of it but we do get an idea just how amazing and complicated and well designed the human body is.

Now, let's look at the blood itself, a fluid that performs many tasks throughout the body. And in order to perform these tasks, it is composed of much more than what we see when a nurse draws blood from a vein or when we cut ourselves. All we see is dark red deoxygenated blood.

When we observe blood we think we see a red liquid, but we are really seeing a fluid, which contains almost 60% plasma, a yellowish sticky substance. And much of that plasma is actually lymph fluid. (Chapter twelve covers how lymph fluid helps protect the human body.) We also see red cells, which make up about 40-45 % of the blood and we also see white cells and platelets, which make up only about one percent of the blood. An added point worth considering - the brain and kidneys control the amount of salt in the blood, and the brain also controls blood pressure.

Red cells contain iron, which receives oxygen from the lungs. White cells, among other things, combat both bacterial and fungal infection, help fight against parasites, and assist in creating an immunity to infection. Platelets are involved in blood clotting. Blood cells are manufactured mostly in bone marrow with some being manufactured in the spleen. Billions of red cells are manufactured in a 24-hour period - red cells are rapidly worn out and they must be replaced constantly.

Platelets, mentioned above, are disc-shaped cells with the task of causing blood to clot when the body receives a wound followed by bleeding. When there is a wound, platelets, which are very sticky, immediately bind to the injured spot (figure 4). They also stick to each other and, combining with fibrin and red blood cells, begin to plug the wound. (Fibrin is an elastic, thread-like protein that helps form the clot.) So fibrin, red blood cells and platelets form a net-like covering, a clot, that stops the flow of blood. While this is happening white blood cells move into the damaged area and fight any infections that might occur. The outside of the clot is exposed to air and forms a scab. Underneath the scab several things happen to remove dead cells, remove the original clot and form new capillaries. Within a few days to a few weeks the wound is healed, the scab falls off and a scar is all that remains.

In this chapter we have only scratched the surface showing the circulatory system and action inside this system. However, we at least understand how complex it is and how its design allows it to perform so many functions in order for the human body to grow and survive.

Blood is a much more complicated liquid than we have described here, but we do get the sense that in a healthy body it does exactly what it is designed to do.

Chapter 2, Figure 4

In this drawing we show how a wound heals. Red blood cells, platelets and strands of fibrin along with other factors all play a part in the healing process.
1. The wound shows a blood vessel has been injured.
2. Blood vessel
3. Red blood cells help form the clot.
4. Platelets form a plug to stop the flow of blood.
5. The scab forms at the surface.
6. Strands of fibrin (yellow) also help form the clot.

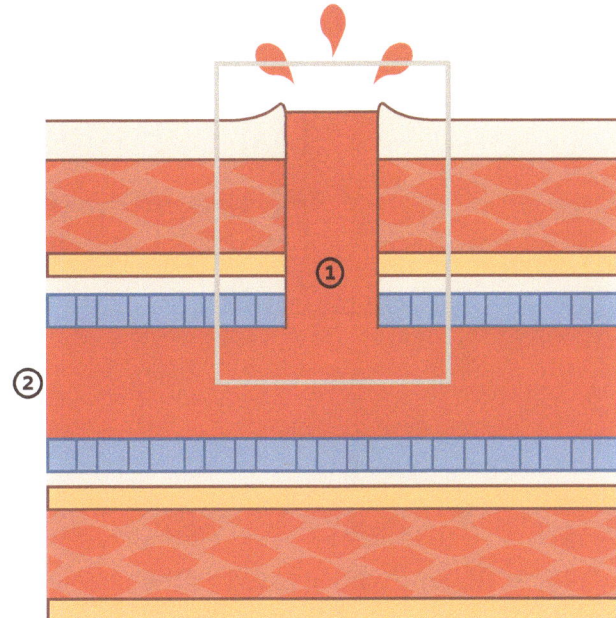
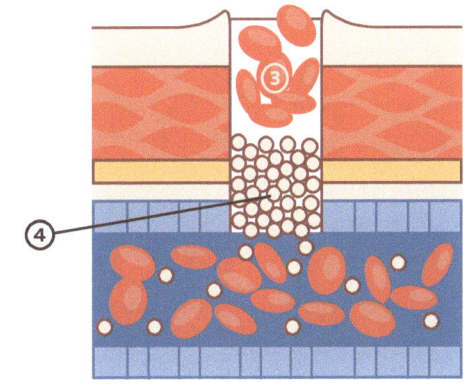
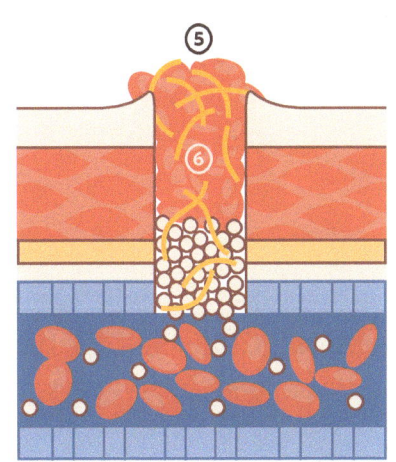

Chapter 3
THE NERVOUS SYSTEM – BRAIN AND SPINAL CORD

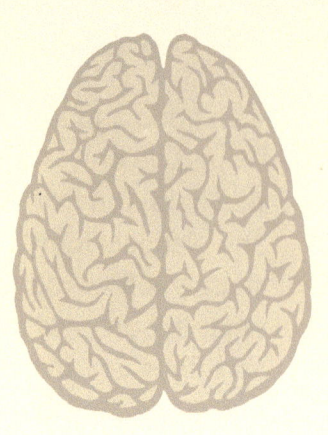

The human central nervous system - the brain and spinal cord - is the most amazing piece of work on this planet. Anyone who has studied the brain and has examined its design knows that it is the most remarkable, mysterious, complicated part of the human body.

Because of the brain we see, hear, taste, smell, feel, maintain balance, breathe, our heart beats, our blood pressure is regulated, we sleep - - this list could go on and on.

The brain lets us think, reason, compute, understand, react, feel emotions - again an almost endless list.

The brain weighs about three and one half pounds but it requires more oxygen-enriched blood per pound than any other part of the human body. And if a brain is deprived of oxygen for just a few minutes - a drowning victim for instance - it will cease functioning and that person will probably be declared brain-dead. At that point the brain cannot be revived although the body can be kept "alive" through life-support systems such as respirators that keep the lungs working.

Blood flows to the brain through two sets of arteries, carotid and vertebral arteries. Carotid arteries supply the cerebral cortex and vertebral arteries supply the brain stem and cerebellum. Carotid arteries are the main arteries in the neck, and vertebral arteries are enclosed in the spinal column. All four join to form the circle of willis at the base of the brain.

The main part of the brain is enclosed in the skull which helps protect it from outside blows. But between the skull and brain are three designed layers of material that further protect the brain (figure 1). The outer layer is called "dura mater", a strong leathery covering. The next layer is a loose spider-like covering called "arachnoid mater". It contains a cushioning liquid known as cerebrospinal fluid. The inside layer, called "pia mater", is filled with blood vessels and is attached directly to the brain. These three protective layers are called meninges. This same system of protection surrounds the spinal cord, which we will investigate later in this chapter.

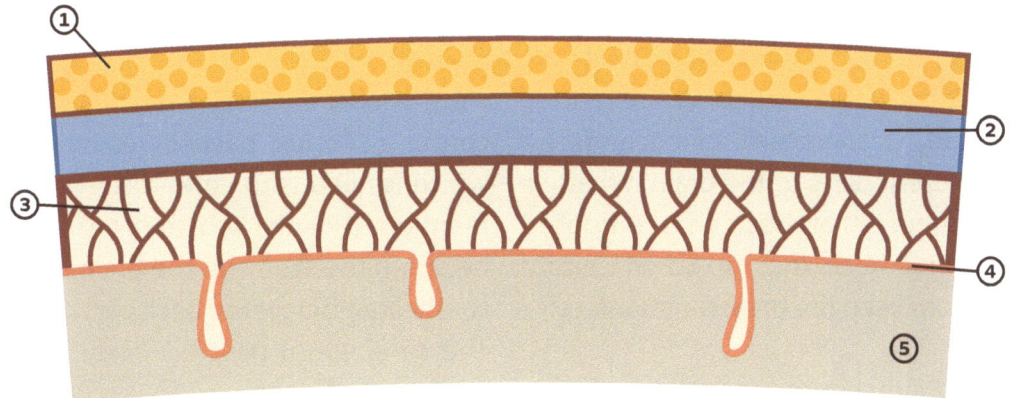

Chapter 3, Figure 1

This is an extremely simplified presentation of the protective coverings of the brain, called meninges. There are three layers of protection - the top layer touches the skull and the bottom layer is attached directly to the brain.

1. Skull
2. Dura mater - A tough, leathery layer
3. Arachnoid mater - A spider-like middle layer filled with a liquid actually produced in the interior of the brain.
4. Pia mater - The third protective layer attached to the brain
5. The top of the brain, the cerebrum

If we remove the skull and protective coverings and look directly at the brain from the top, we see in figure two the main part of the brain, the cerebrum. Notice that the cerebrum is divided into two hemispheres or halves, which reveal a wrinkled gathering of hills and valleys. (The front of the brain is at the top of the drawing.) The hill or ridge is called a gyrus and the valley is called a sulcus. (Plurals are sulci and gyri.) The longitudinal fissure is the dividing valley between the two hemispheres.

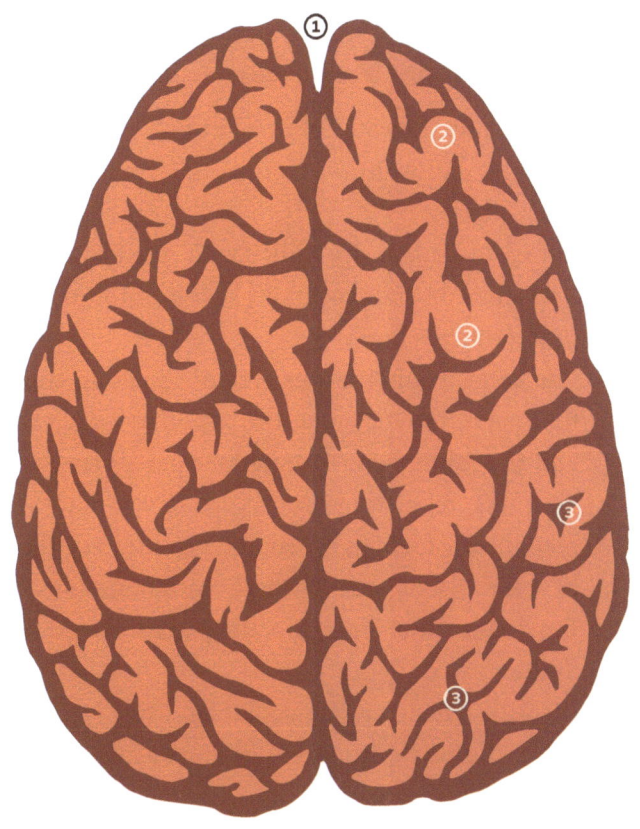

Chapter 3, Figure 2

Looking at the brain from above with the skull and protective covering removed. We are looking at the cerebrum.

1. Longitudinal fissure divides the brain into two hemispheres.
2. Hills or ridges - A hill is called a gyrus.
3. Valleys - A valley is called a sulcus.

We have heard about the brain's grey matter and white matter. Grey matter is the outside surface, which contains nerve cell bodies. White matter, the interior part, is composed of nerve fibers connecting different parts of the brain. What we need to understand is that the brain is composed of millions of nerve fibers that can communicate with each other to keep the brain functioning.

The two hemispheres are connected deep inside the brain by the corpus callosum, a thick band of nerve fibers (figure 3). Continuing to look at figure three, we see the interior of the brain while we are also viewing the right cerebrum. The corpus callosum partially covers the ventricle, which is filled with cerebrospinal fluid, the fluid mentioned earlier which fills the middle lining between the skull and the brain. We also see the cerebellum below the back part of the cerebrum. The cerebellum controls body movement and maintains balance.

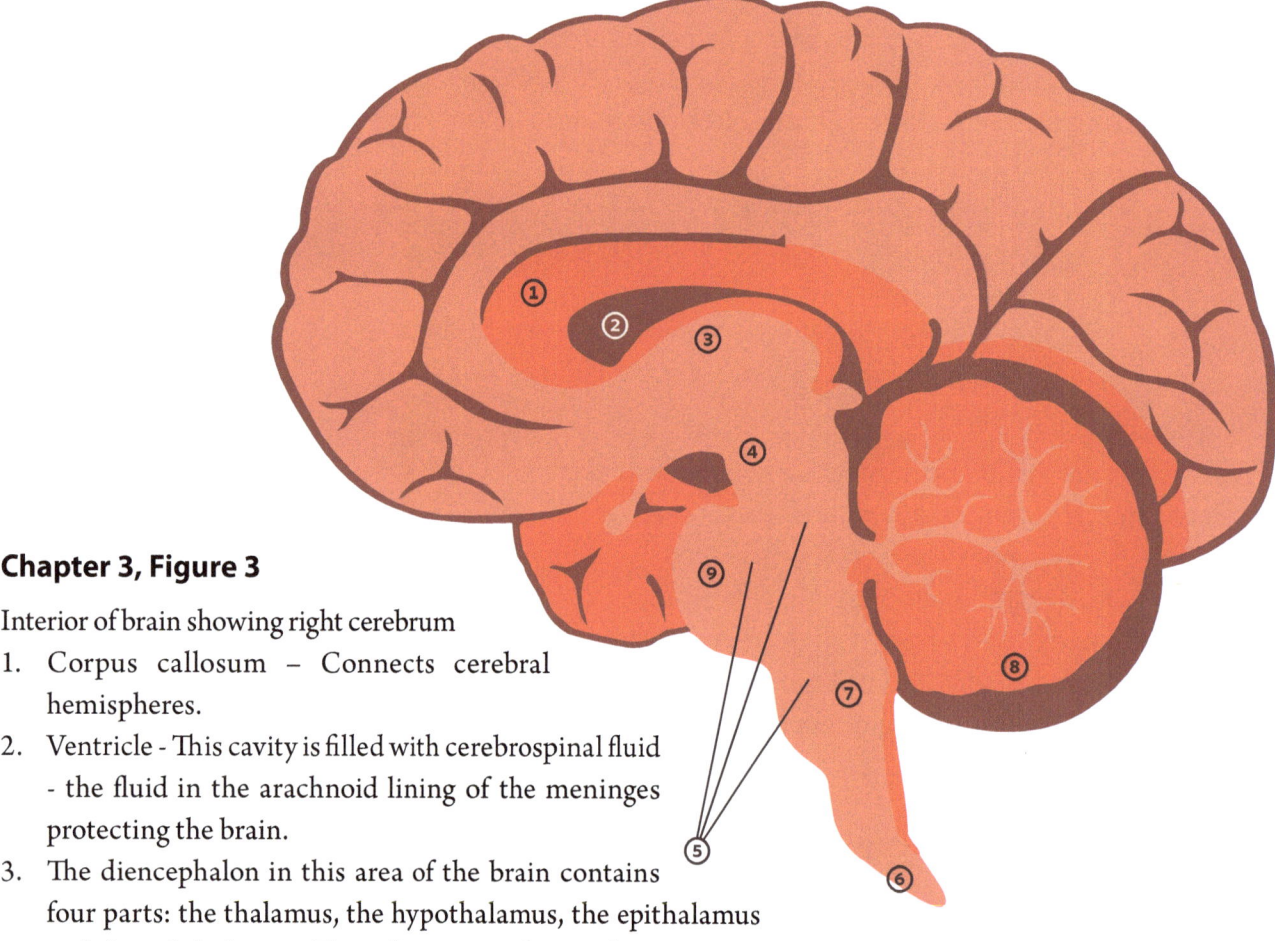

Chapter 3, Figure 3

Interior of brain showing right cerebrum
1. Corpus callosum – Connects cerebral hemispheres.
2. Ventricle - This cavity is filled with cerebrospinal fluid - the fluid in the arachnoid lining of the meninges protecting the brain.
3. The diencephalon in this area of the brain contains four parts: the thalamus, the hypothalamus, the epithalamus and the subthalamus. These four parts play a role in recognizing pain, temperature, emotions, memory, alertness, regulating certain muscles, blood pressure, regulating body temperature, feelings of hunger, recognizing feelings of fear and pleasure, regulating the sleep-wake cycle and controlling movement.
4. Midbrain - Connects forebrain with hindbrain
5. Brainstem - Includes midbrain, pons and medulla oblongata
6. Spinal cord
7. Medulla oblongata - Controls blood supply, heartbeat and breathing
8. Cerebellum - Body movement and balance are centered here.
9. Pons

Because the brain is so complicated we cannot gain a complete understanding of its structure and how it accomplishes so many tasks simultaneously. The medical student spends countless hours on these topics. We will just get a brief look at some of the brain's major functions.

So much happens, yet we pay little attention.

Located below the corpus callosum is an area called the diencephalon. In this area a part called the thalamus relays information from sense organs to another part of the brain, the cerebral cortex. Also in the diencephalon is the hypothalamus which deals with hunger and thirst drives and helps control body temperature and the water-salt balance in the blood.

Now we can turn to memory. We know we can memorize something and keep it in our memory almost forever. But our brain actually has two memory features - short term and long term. We look for a telephone number, remember it long enough to dial the number, then we probably forget it. Somehow our brain knows not to keep that number in its long-term memory. But we want to memorize a poem and we want to retain that poem in our memory. Our brain knows we want to retain that poem and it is stored in long-term memory.

Taste buds in our mouth, especially on our tongue, send signals to our brain that we taste something. Immediately our brain deciphers the signal and tells us what we taste. Even if we are blindfolded and someone puts food in our mouth, our brain deciphers the taste and we know immediately what is in our mouth. The point is - even though we taste it immediately, the communication between the taste buds and the brain had to happen before we recognized what we were tasting.

A mosquito bites us on the arm and before long our skin begins to itch. But nerves in our arm send signals to the brain and the brain tells us it is a mosquito bite. At that same location we can cut our arm and the brain will let us know it is a cut - not a mosquito bite.

If a person decides to raise his arm, his brain sends a signal to the appropriate muscles and he raises his arm, probably not even thinking about the communication between his brain and his arm. The desire to raise his arm is recognized in one part of the brain, but that desire must be communicated to another part of the brain before the arm muscle reacts. A stroke victim with part of his body paralyzed has actually experienced brain damage and that damage keeps the victim from moving the part of his body affected by the stroke. If he wants to move his arm but the part of the brain that would cause his arm to move is damaged he cannot move his arm. The muscles are still there but they cannot move because the brain cannot send the signal to the arm muscle.

Our lungs contain no muscle, yet we breathe constantly and think nothing of it. But the brain actually keeps us breathing by sending a signal to our rib muscles and diaphragm to expand. The lungs expand with the ribs and diaphragm and fill with air. Then the brain sends another signal to the ribs and diaphragm and they voluntarily contract, squeezing our lungs, forcing air out. This is the breathing cycle to which we pay almost no attention.

Sound waves travel through the outer, middle and inner ear, are converted into electrical impulses, and then are transported to the brain. The brain then interprets those signals and lets us know what we have heard. And this process is instantaneous.

Our eyes focus on an object and the retina receives that image. The optic nerve receives that information from the retina and sends that image to the back part of our brain, where it is interpreted. Only then do we know what we have seen and, again it is instantaneous.

Although these are just some examples of brain activity, we can begin to understand how amazingly complicated our brain is.

For instance, the following is probably "over our heads", but read it anyway to get an idea how much science has learned about the central nervous system.

The brain is filled with nerve cells called neurons, and connections between neurons are called synapses. There are about one thousand trillion synapses in our brain. There are about one hundred billion neurons in our nervous system and about fifty trillion neuroglia - cells that support and protect neurons.

The more we read, the more we understand why the medical student must spend countless hours working and studying in order to become a member of the medical profession.

THE SPINAL CORD

Back in the first part of this chapter we learned that the brain and spinal cord make up the central nervous system. We now have a basic grasp of the brain, so we know how important and complicated it is. Since the brain and spinal cord are connected we can now look at why and how that connection is so important.

Referring to figure 3, connections between the main parts of the brain and spinal cord are shown. The midbrain connects the front and back parts of the brain. The pons is actually part of the brainstem. It contains many nerve tracts, which are passageways for signals to and from the brain.

The medulla oblongata is the lowest part of the brainstem and controls blood supply, heartbeat and breathing. So the part we call the brainstem is composed of the midbrain, pons and medulla oblongata.

The spinal cord extends from the medulla oblongata down through the spinal column (figure 4) and is full of nerve tissue, which carries messages to and from the brain. The spinal cord has the same protective layers as the brain - the dura mater, the arachnoid mater and the pia mater, and these coverings are also called meninges. The spinal cord is enclosed in the vertebra of the spinal column so it is protected by bone, cartilage and three layers of special covering.

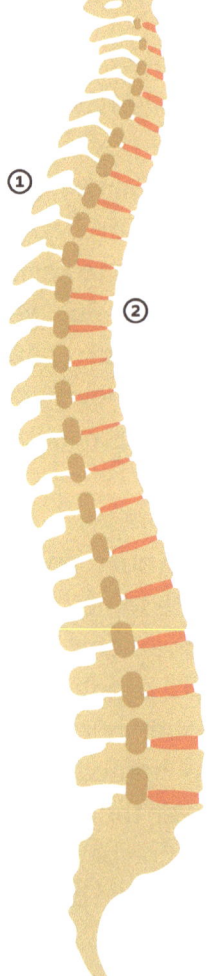

Chapter 3, Figure 4

The spinal column plays a major role in almost all body movement as well as body stability. However in this chapter we only want to show its relationship to the spinal cord. Chapter ten gives an in-depth presentation of the spinal column and the body's bone structure.
1. The spinal cord is located inside the spinal column and is protected by meninges, bone and cartilage.
2. Spinal column

Tracts in the spinal cord are composed of extensions of neurons, called axons, which actually send messages both ways. Ascending tracts carry messages up to the brain and descending tracts carry messages down and out to all parts of the human body. If the brain receives a message that your hand is about to be burned because it is too close to a hot object, it will immediately send a message to pull your hand back. This can happen because nerves stretch out from the spinal column to all parts and surfaces of the body.

The brain was involved in this example of nerve reaction. But when a doctor taps a person's knee and causes a reaction showing the leg from the knee down jumping forward, this is an example of involuntary reaction. The nerve at the knee sent a signal to the spinal cord, which in turn caused the leg to move without any contact message from the brain.

We have learned that the brain and spinal cord make up the central nervous system. All of the nerves that reach out from the spinal column compose the peripheral nervous system, which is composed of millions of nerves that must connect back to the central nervous system.

As you read through the rest of this book you will receive more in-depth explanations showing how the peripheral and central nervous systems work with all parts of our bodies. But for now, we can recognize that much design and planning went into what the human body is - an amazing "machine" that has so many parts that must work together constantly to keep the body alive and well.

Chapter 4
LUNGS

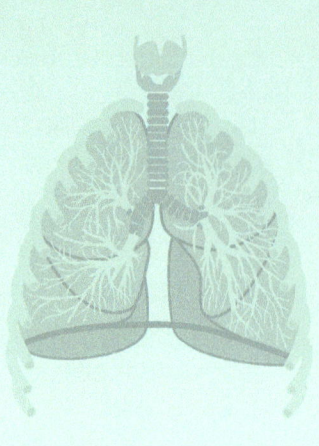

Our lungs are two organs which lie on either side of the heart. Their upper tip (apex) extends behind each collarbone. The bottom of each lung also rests on the diaphragm. They are two of the largest organs in the body and are protected by the ribs (figure 1).

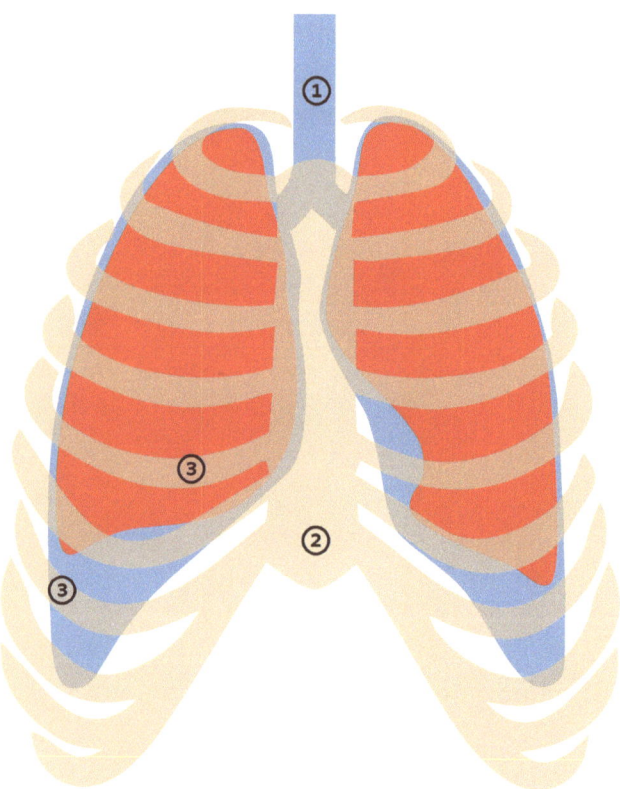

Chapter 4, Figure 1

A view of the lungs, ribs and sternum
1. Trachea - Divides into two bronchi, which enter each lung. These bronchi divide into smaller and smaller vessels and move air to the entire lung.
2. Sternum (breastbone)
3. The lungs are surrounded by the ribs which are attached to the spinal column in the back and to the sternum in the front. This design protects these two very important organs.

Although the lungs appear to be two large masses, they are actually divided into lobes, three in each lung, but the left lung is slightly smaller. There is reason for the difference - the left lung gives up some space to the heart, which accounts for the difference in size. It still has three lobes but one is slightly smaller and is known as the lingula. But each lobe operates independently because all of the air passages and blood vessels divide as they first enter the lungs and efficiently serve each lobe.

The lobes are held firmly together by a lining called the visceral pleura. Another lining inside the ribs and on top of the diaphragm is separated from the visceral pleura by a very narrow space filled with a slippery fluid, which allows the two linings to slide over each other quite easily. This feature allows the lungs to maintain the same contour as the ribs and diaphragm as they expand and contract during breathing.

The diaphragm, mentioned above is a sheet of muscle that separates the abdominal and thoracic cavities. It is essential for breathing as explained later in this chapter. The diaphragm appears as a solid sheet of muscle but it actually has three openings for the esophagus, the aorta and one of the main veins taking blood back to the heart (figure 2).

The function of the lungs is to receive fresh air breathed in through the nose or mouth, supply oxygen to all parts of the body and receive carbon dioxide from blood that has come from all parts of the body. The lungs then expel that gas. That sounds simple enough but the process is anything but simple.

Fresh air enters the lungs through the trachea (wind pipe) with its upper end beginning behind the tongue at the larynx. Air moves down the trachea to bronchi, which open into each lung. The bronchi are formed at the end of the trachea and enter each lung at about the mid-point, called the hilum (figure 2).

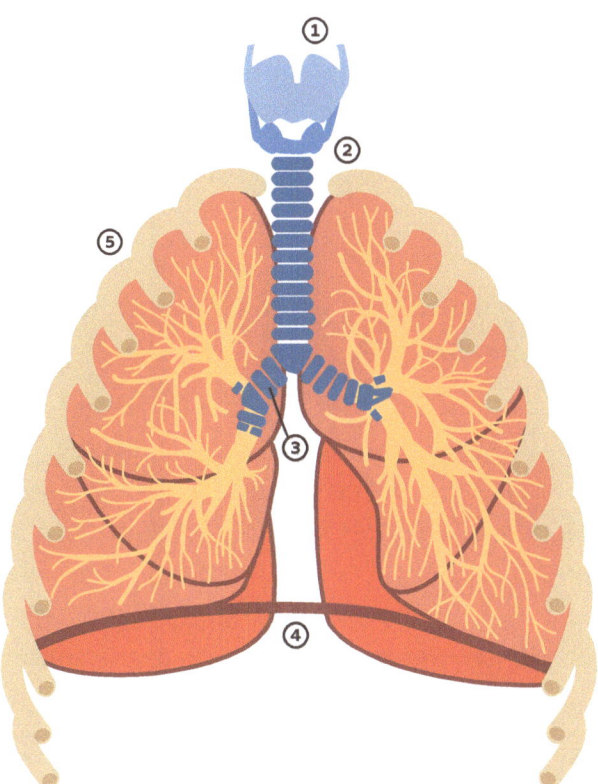

Chapter 4, Figure 2

Pictured are the trachea, each bronchus, a cut-away view of the interior of each lung and a partial view of the ribs.

1. Larynx
2. Trachea - Transports air from the mouth/nose to the lungs.
3. Bronchus - A bronchus enters each lung making a passageway for air. The point at which each bronchus branches off and enters the lung is called the hilum
4. Diaphragm - The diaphragm is shown supporting each lung. The diaphragm and ribs receive signals from the brain to expand causing the lungs also to expand and pull in air (they inhale). The brain then sends signals to the diaphragm and ribs to relax causing the lungs to exhale. The lungs have no muscles of their own.
5. The ribs surround the lungs providing protection.

The interior of each lung is designed to provide as much surface area as possible for the exchange of oxygen and carbon dioxide. Deoxygenated blood is pumped from the heart's right ventricle to each lung. When the exchange of oxygen and carbon dioxide has taken place, veins transport the oxygen rich blood back to the left atrium of the heart, then to the left ventricle from where it is pumped through the aorta to the entire body (chapter one).

But when we examine the interior of the lungs in order to "watch" the exchange of gases, we discover an amazing design with different parts that have different functions and they all had to come together at the same time so that the lungs and heart and brain can work together to keep the body alive and healthy. When air reaches the lungs through the bronchi, it moves through smaller and smaller vessels until it reaches the alveoli (called alveolar sacs) where the actual exchange takes place. There are millions of alveoli in the lungs (figure 3).

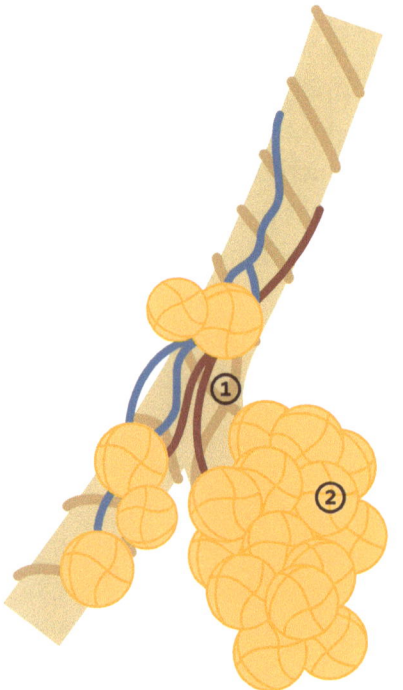

Chapter 4, Figure 3

A view of alveolar sacs, microscopic air-exchangers in the lungs. Each bronchus divides into smaller and smaller passages. The smallest are called bronchioles.
1. The bronchioles send air to the alveolar sacs where capillaries relieve blood of carbon dioxide and move oxygen into the blood stream. Each alveolar sac is covered with arteries, veins and capillaries. So inhaled air contains needed oxygen and exhaled air contains carbon dioxide with the exchange taking place in the microscopic alveolar sacs.
2. Alveolar sacs - where the exchange of oxygen and carbon dioxide takes place.

The alveolar sacs are surrounded by blood vessels, both veins and arteries, and the alveoli, through capillaries connecting the veins and arteries, take the carbon dioxide from the arteries and deposit oxygen in the veins. The veins take the oxygenated blood to the left side of the heart as described above. The carbon dioxide is exhaled through the same alveoli that supply inhaled air.

Three separate sets of vessels enter and leave the lungs in order to accomplish this process. The bronchi transport air in and out of the lungs. But the arteries and veins are separate systems. One other system also transports fluid from the lungs - part of the lymphatic system, which is covered in chapter 12.

To get a grasp on what is happening in our lungs take three or four deep breaths. During the inhaling and exhaling of each breath, the entire process previously described has taken place. We normally breathe in and out about 15 times each minute and the exchange of oxygen and carbon dioxide takes place in our lungs every time we inhale and exhale.

We breathe two different ways.

If you took the recommendation above and drew three or four deep breaths, you controlled your breathing. But as you relax and forget about breathing, your brain, diaphragm and ribs still "see to it" that your lungs continue to function as they should/must. For example when we are asleep the process takes care of itself, with the brain, diaphragm and ribs still doing what they are designed to do.

The respiratory center in the hindbrain has two regions - an inspiratory and expiratory center. The hindbrain, for our purposes in this chapter, is simply a part of the brain toward the back.

The inspiratory center sends nerve impulses to the rib muscles and diaphragm telling the ribs to expand and telling the regions of diaphragm, which support the lungs to contract downward causing the lungs to expand. This expansion allows air to enter the lungs. Whether we are controlling our breathing or letting the brain control it by sending signals to the rib muscles and diaphragm, the process inside the lungs is the same.

As our lungs fill with air, the walls contain stretch receptors, which send signals to the inspiratory center in our brain telling it to stop the expansion of the ribs and contraction of the diaphragm. At the same time the expiratory center tells the diaphragm and ribs to relax, causing the lungs to decrease in volume so that air is exhaled. When we are asleep this procedure keeps occurring and when we are awake and not controlling our breathing, the same procedure takes place.

When we are exercising or in some way experiencing stress, our brain senses that we need more oxygen and causes more rapid breathing and a speeded up heart rate in order to get more oxygen from our lungs and send it to the parts of our body that need a greater supply.

So now we know something about how our lungs work. We know that tiny sacks in our lungs, alveoli, are covered with veins and arteries and capillaries. These alveolar sacks take carbon dioxide from our blood and at the same time put fresh oxygen into our blood. This process takes place every time we inhale and exhale. We also know that a part of our brain constantly sends signals to our rib muscles and diaphragm to keep expanding and relaxing allowing our lungs to inhale and exhale about 15 times each minute. We know that our heart pumps deoxygenated blood to our lungs and receives oxygenated blood from our lungs and sends that blood to all parts or our body. If the arteries that take blood from the heart to the lungs, and if the veins that take blood from the lungs to the heart, were not in place, the lungs would be worthless. If our brain did not send signals to our ribs and diaphragm, we would not breathe. The connecting vessels between our heart and lungs and the nerves that send signals from our brain to our ribs, diaphragm and heart are designed to keep our body alive and healthy.

Chapter 5

DIGESTIVE SYSTEM

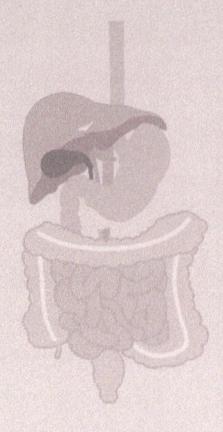

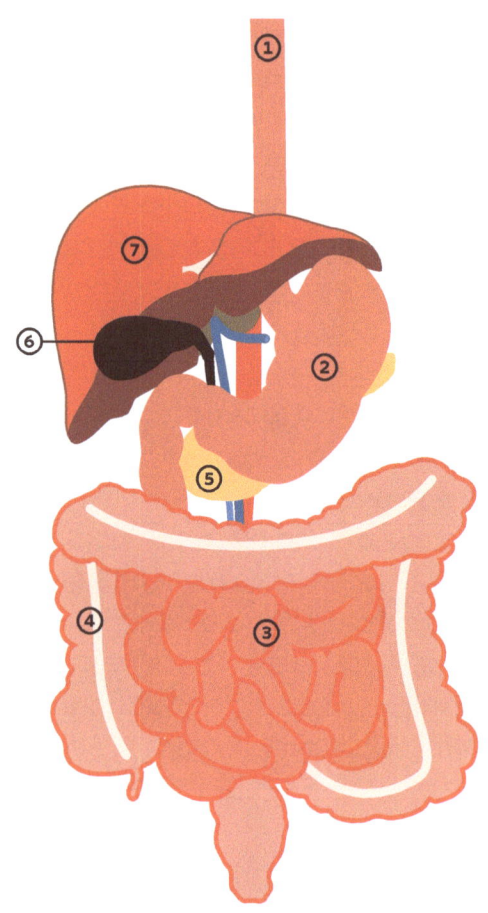

Chapter 5, Figure 1

The entire digestive system.
1. Esophagus
2. Stomach
3. Small intestines
4. Large intestines
5. Pancreas
6. Gall bladder
7. Liver

In order to understand how food is digested and absorbed in our body, we must first learn how much of our body is involved in the process. Many different things happen - more than we might originally have thought.

Here is what most of us think about how we digest food. We take a few bites of food, chew it for a few seconds, swallow it and keep taking more helpings until we have consumed a meal. We know that the food entered our stomach, got worked over pretty well with the necessary nutrients being absorbed somehow into our blood stream and taken to the parts of our body that need the nutrients. The food that is not absorbed moves on to our intestines and is eventually expelled from our body. But if we take the time to examine the digestive process we are amazed at how this entire system actually works. So let's take a few minutes to get a better look at everything that happens to keep us alive and well as we consume two or three meals every day.

Starting at the "top" figure one begins with the esophagus and continues with the stomach, small intestines, liver, gall bladder, pancreas and large intestines - all involved in some way in the digestion of food. And there is another part to the story. Once the food is digested the resulting nutrients must be moved to the parts of the body that use it to survive, grow, repair and replenish. This means the blood stream plays a major role.

And finally, the nervous system is also in the picture. Nerves fan out throughout the body but they also, in some cases, converge in clusters. This happens in the GI tract where the clusters become involuntary mini-controllers of digestion.

The food itself is restricted to the mouth, esophagus, stomach and intestines. These parts make up the part of the gastrointestinal tract (GI tract), which actually handles the food. This part of the GI tract is like a long tube that begins in the mouth and terminates at the lower end of the intestines. In an adult this tract is about thirty feet long. The other parts mentioned above (the liver, gall bladder and pancreas) never "see" the actual food but without them the body could not survive.

THE GASTROINTESTINAL TRACT

We will look at the GI tract first and we'll begin by observing how food is actually swallowed. We bite off a piece of food and we chew it so that we can swallow it. (The amount we swallow is called a bolus) When we finish chewing, the tongue pushes up against the hard palate (the roof at the front of our mouth) and forces the food to the back of our mouth. Now the food (bolus) forces the soft palate (also considered part of the roof) to move up and close off the nasal cavity. The epiglottis closes off the trachea (wind pipe) (figure 2), and the food is forced into the esophagus through the throat. The bolus is now on its way to the stomach. When the epiglottis closes off the trachea it keeps us from choking. Otherwise food would have headed for our lungs.

But while the food was in our mouth taste buds and salivary glands became active so we actually tasted the food, and saliva was secreted, mixed with food, and began aiding the digestive process. In the esophagus, peristalsis moves the food to the stomach. For some reason we probably think digestion takes place mostly in the stomach. But the stomach is a muscular bag situated below the diaphragm and its main function is to churn the food, mix it with gastric juices forming a thick fluid known as chyme. This is the beginning of the digesting of fats and proteins.

Before we follow the food any farther, let's take a closer look at the stomach. This muscular bag is in a somewhat collapsed state when it contains no food. But it can expand to hold a full meal easily. The stomach secretes hydrochloric acid and one of its functions is to kill most types of bacteria and other micro-organisms. Gastrin is a type of hormone, which helps to stimulate the production of hydrochloric acid; gastrin is then absorbed into the body of the stomach from where it moves to certain blood cells. The stomach's contents are so acidic that they could dissolve a razor blade, which obviously means the walls of the stomach actually could be digested. To prevent this unpleasant thought the walls of the stomach are protected by a layer of alkaline mucus and the cells lining the stomach are replaced so rapidly that the stomach has a complete new lining every three days.

Going back to the food that enters the stomach, we just watched the progress of one bite. The stomach ordinarily holds an entire meal - it can hold a big meal because it is a muscular bag that can stretch well beyond its size when it is empty. With the mixing and churning that takes place, the stomach is ready to empty its contents into the small intestines. But the stomach empties its contents very slowly because the small intestines, in a round about way, limit how much food can enter. The small intestines secrete into the blood a hormone known as CCK. This hormone keeps the stomach from emptying its contents too quickly into the small intestines, which means that it takes about four hours for the small intestines to digest a meal.

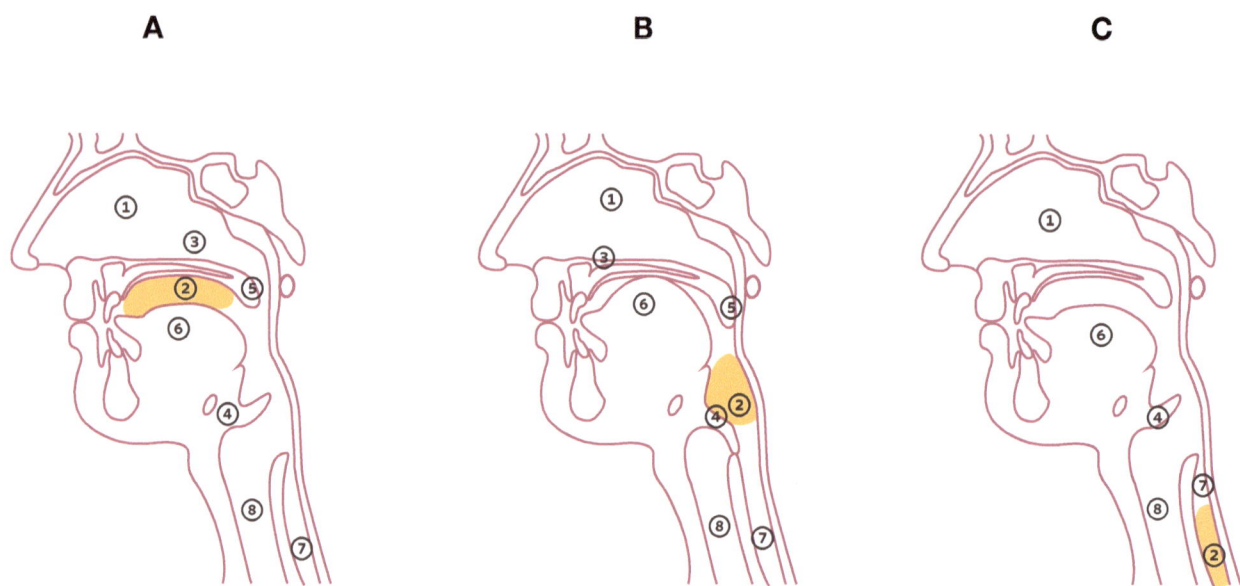

Chapter 5, Figure 2

Bolus and epiglottis - These three drawings show food being chewed and swallowed. The interesting part is following three distinct maneuvers as the food (bolus) moves on its journey. In figure "A", the food is the tan mass between the tongue and hard palate. Notice that while the food is still in the mouth the opening remains from the nasal cavity down through the trachea, which means we breathe safely while we are chewing. In figure "B" the tongue has pushed the bolus against the soft palate closing off the nasal cavity and at the same time the epiglottis has closed off the trachea. In figure "C" the bolus has moved into the esophagus, the soft palate is no longer closing off the nasal cavity and the epiglottis is in the open position, so we breathe until we swallow again. Peristalsis now moves the food to the stomach. The hard palate, tongue, soft palate, epiglotttis and esophagus are designed to work together to allow us to chew and swallow our food safely.

In "B" above, the epiglottis closes off the trachea (wind pipe) when the food is being swallowed.
The trachea is open except when food is being swallowed.
In "C" above, the epiglottis and soft palate are open allowing for normal breathing.

1. Nasal cavity
2. Bolus (food)
3. Hard palate
4. Epiglottis
5. Soft palate
6. Tongue
7. Esophagus
8. Trachea (wind pipe)

The small intestines become a kind of chemistry lab during the digestive process. We don't need to try to understand all of the chemistry involved in the digestion of food. That is for the medical student to deal with. But we can get some idea how complicated the process is. For instance, the digesting of fats and proteins begins in the stomach but it is completed in the small intestines.

The small intestines contain tiny finger-like projections, called villi, (figure 3) which give the walls of the intestines enormous surface area. In the villi are arteries, veins and lymph vessels, and all of them play a role in the digestion of food. Since the chyme was thoroughly mixed in the stomach it is now a fluid that can easily flow into the tiny spaces between the villi. (The small intestines are actually divided into three parts. The first is the duodenum which is connected to the stomach, the second is the jejunum which is the longest part of the small intestine and the third is the ileum which empties into the large intestines which carry waste out of the body).

So from the time the chyme has entered the small intestines until it enters the large intestines, it is a liquid which passes very slowly through the intestines, and its nutrients have been absorbed by the blood vessels in the villi. Those vessels transport all of the nutrients to the liver where they are processed further before being transported to all parts of the body.

THE LIVER

The liver is the largest gland in the human body. It weighs about three pounds, is located above the stomach and is protected by the right rib cage (figure 1). It is divided into two main parts - the left and right lobes. The left lobe is actually divided into smaller lobes but we don't need to deal with all of the "architecture" of this gland. What we do want to understand, to some degree, is how the liver functions and also how many different actions are designed into this amazing three pound wedged-shaped mass in the human body.

One function of the liver is to produce bile, which breaks down fats. However the bile is not stored in the liver, instead it is actually stored in the gall bladder (figure 1), and empties in small amounts into the small intestine, the duodenum, where it breaks down fats. Bile is a liquid composed of different fluids, which eventually aid in the digestion of food. (When fats are present the stomach sends a signal to the gall bladder to secrete bile into the small intestines).

To place everything in proper perspective we must understand that the liver is one of the most complex organs in the human body, performing at least 500 chemical reactions, not all of them directly related to the digestion of food. For instance, the liver breaks down worn out red blood cells, a process which it partially uses to make bile. It also makes a protein, which affects blood

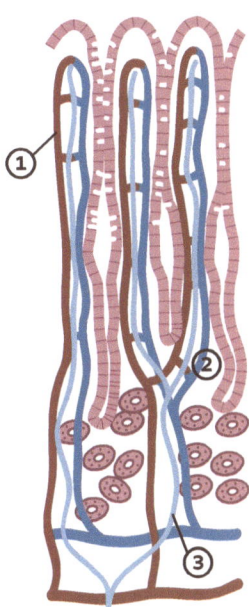

Chapter 5, Figure 3

Villi in the small intestines: These finger-like projections in the small intestines provide much added space for the digestive process. The chyme is in liquid form and slowly passes by/through the spaces between the villi. The blood vessels in the villi absorb the nutrients in the chyme and transport those nutrients to the liver.
1. Artery
2. Vein
3. Lymph vessel (Chapter 12 covers lymphatics)

clotting. And the liver manufactures cholesterol and stores minerals and certain vitamins, such as A, B12 and D.

Another function of the liver is to break down toxic substances, such as alcohol. A person who consumes great amounts of alcohol runs the risk of cirrhosis, which is a hardening of liver cells and eventually causes the liver to quit functioning. At this point the person must either receive a liver transplant or face ultimate death.

Through another interesting process the liver "balances" fats and carbohydrates. If there are not enough carbohydrates in the diet to produce the proper amount of energy, the liver breaks down stored fat into chemicals that produce heat and energy.

Finally, we need to understand that blood left the small intestines "loaded" with nutrients and was transported to the liver through the Hepatic Portal Vein. Then the liver worked on these nutrients through many chemical reactions. After the liver has done what it is designed to do the nutrients reenter the blood stream and are transported to the heart and then to all parts of the body. The food that entered the small intestines but was not absorbed as nutrients is now waste and moves on to the large intestines and finally out of the body. The large intestines also release water into the blood system, which keeps the waste in a mostly solid form.

THE GALL BLADDER AND PANCREAS

As mentioned earlier, the gall bladder stores bile, which is manufactured in the liver. Bile is used in the digestive process to break down fats and it is also essential in the absorption of vitamins D and E. Bile is secreted into the small intestines (into the duodenum) when fats are present in the stomach. (Figure one shows the location of the gall bladder and pancreas.) The pancreas produces both hormones and enzymes and secretes them into the duodenum to help digest food. It also produces insulin. The pancreas plays a very essential part of the entire digestive process. If the Pancreas is removed, medication must be available to continue supplying the body with the fluids manufactured by, or stored in the pancreas. The gall bladder also can be surgically removed if it becomes inoperative or inflamed mostly because of gallstones. Medication helps the body adjust to the absence of the gall bladder. (The body can live without a gall bladder and pancreas but it cannot survive without a liver.)

Obviously we have not covered all of the intricate details as we try to understand the digestive system, but we do get the idea that the design of this system is amazing, to say the least.

Chapter 6
KIDNEYS

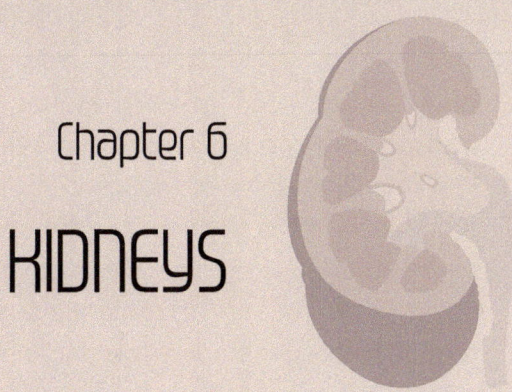

Two kidneys are situated in the back of the abdomen on either side of the spine and are partially protected by the lower part of the rib cage (figure 1). They are relatively small in size but their task is to purify the blood in our bodies. Each kidney has been described as about the size of a deck of playing cards. However, the body's entire blood supply passes through the kidneys many times each twenty four hour period.

Blood reaches the kidneys directly from the aorta through renal arteries, and when it leaves the kidneys, it returns to the heart through the main vein coming from the lower part of our body. Without kidneys, our bodies could not live - unless a person is on dialysis, a system which uses an outside machine to purify blood - and without our heart, the kidneys could not function. The large volume of blood that passes through our kidneys each day is testimony to how important our kidneys are.

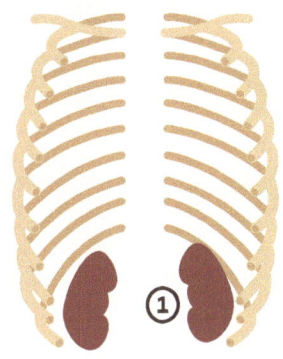

Chapter 6, Figure 1

Each kidney is only about the size of a deck of playing cards but the body's entire blood supply passes through the kidneys many times each twenty four period.
1. Kidneys

When blood reaches each kidney, the artery transporting the blood into the kidney divides into smaller and smaller vessels until it reaches the functional unit in the kidney - the nephron. And we need to know - there are approximately one million nephrons in each kidney (figure 2). The nephrons are located in the part of the kidney called the Renal Cortex, the outer-most region of the kidney. The middle region is called the Renal Medulla and the inner region is known as the Renal Pelvis (figure 3). Inside the Renal Medulla are units called Renal Pyramids. The Renal Medulla and Renal Pelvis serve to collect urine and finally send it to the bladder.

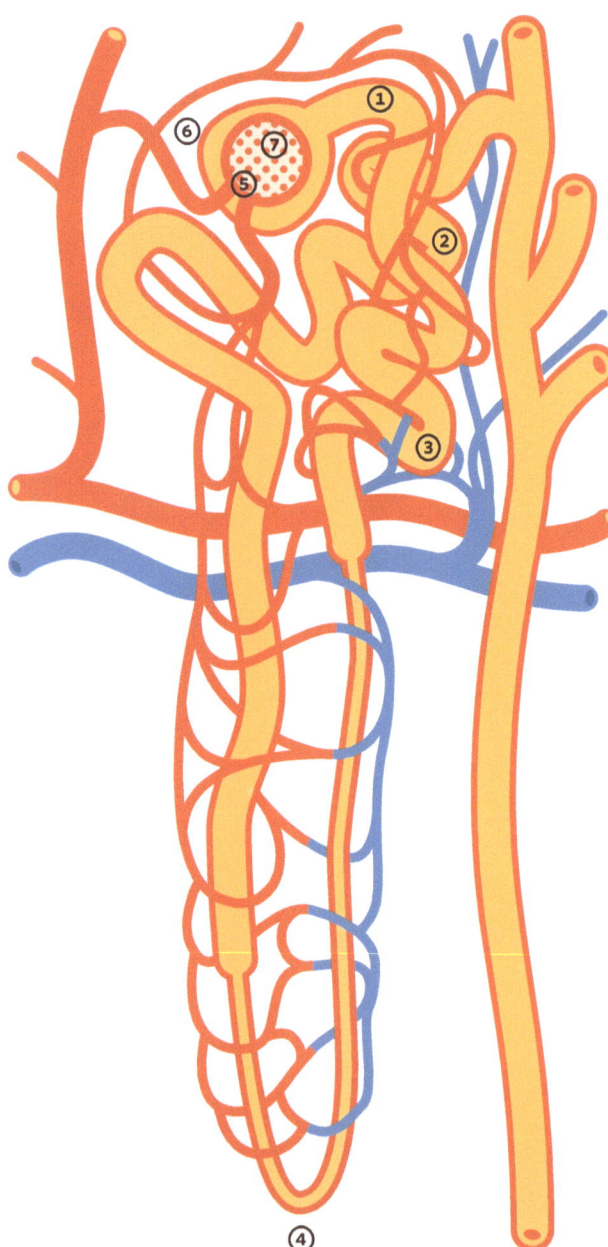

Chapter 6, Figure 2

Nephrons are microscopic parts of the kidney. There are approximately one million in each kidney and it is here that filtering takes place.

1. Proximal convoluted tube – Follow this tube from this point down through the Loop of Henle and up to the point where it drains into the collecting duct. This duct begins moving urine to the bladder. When the tube begins its journey up from the Loop of Henle, it becomes the distal convoluted tube. This is the route the filtrate follows in about one million of these nephrons in each kidney.
2. Distal convoluted tube
3. Proximal convoluted tube
4. Loop of Henle
5. Plasma enters the glomerulus here.
6. Glomerular capsule - also called the helmet.
7. Glomerulus - the filter

Kidneys

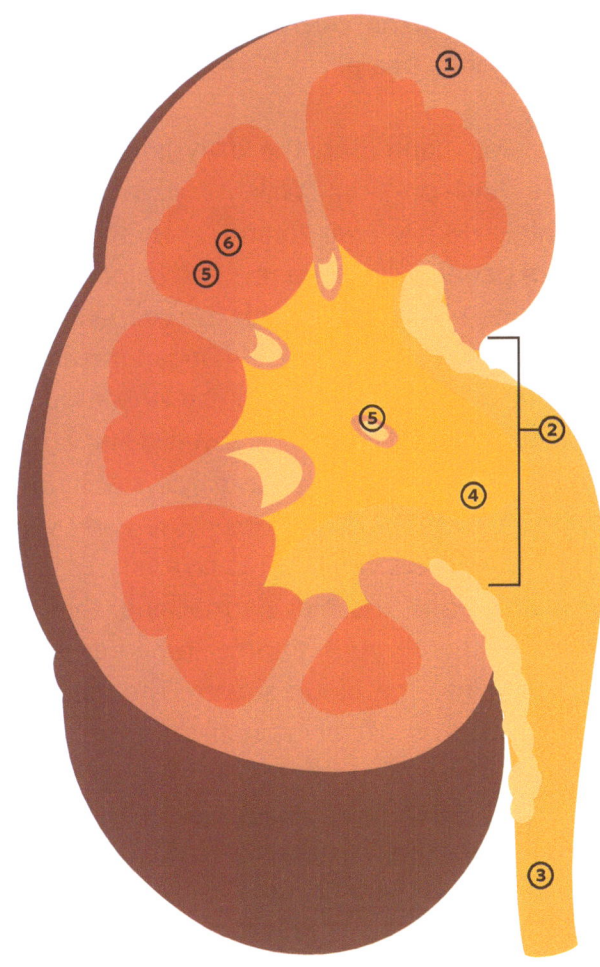

Chapter 6, Figure 3

Cut away view of the kidney
1. Renal cortex - Nephrons are located here.
2. Hilus - Where the ureter begins and where blood vessels enter and leave the kidneys.
3. Ureter - carries urine to the bladder.
4. Renal pelvis
5. Renal medulla and renal pelvis collect urine and send it to the bladder.
6. Renal medulla

Our goal in this chapter is to gain an understanding of how the kidneys receive toxic blood, purify it and send it back to the heart ready to take oxygen to the entire body. And the nephrons play the major role in the purification process.

Nephrons have been described as "miraculous miniature filtration plants of tubes and capillaries". Look again at figure two. You are viewing a greatly enlarged illustration of a nephron which is actually a microscopic unit inside the kidney - about a million of them inside each kidney, as noted earlier. This gives us an idea how amazing the nephrons are and how efficient they must be in order to filter the amount of blood they handle every hour.

In the illustration of the nephron, (figure 2) you notice at the top is a spherical object called a glomerulus. This part of the nephron is the filtration unit containing a tangled web of capillaries, which filter the blood plasma that has entered the kidney. The cover of the glomerulus is called the glomerular capsule. (It is also called the helmet.) The filtrate from the capillaries enters the inside part of the glomerular capsule, then into the tube called the proximal convoluted tube. It is now known as glomerular filtrate and is the beginning of urine, but most of the water in the urine is reabsorbed into the blood as it passes through the proximal convoluted tube, the Loop of Henle and the distal convoluted tube. You notice that the distal convoluted tube is connected to the collecting duct which transports the remaining waste to the Renal Medulla, the Renal Pelvis and finally to the bladder. (Of the forty or more gallons of the glomerular filtrate that pass through the glomeruli daily, only one to two quarts actually become urine.)

Trying to understand the function of the Distal and Proximal tubes and the Loop of Henle is a task for the medical student. But we can recognize this - along with the water, some solutes are also reabsorbed including chloride ions, sodium ions, bicarbonate ions, glucose, potassium ions, urea, uric acid and proteins. When we realize how complicated the work of these microscopic nephrons is, and that there are approximately one million in each kidney, we begin to understand just how amazing the design of our kidneys really is.

The above description of the work of the nephron is very basic. The important points to remember are that the microscopic nephrons must function as they are designed to function, which means that all of the blood in our bodies has passed through the kidneys many times each 24 hour period. And the impurities in our blood have been filtered out and the "cleaned" blood returns to the body's blood stream.

Earlier we mentioned the dialysis machine. In 1941, a Dutch physician, Willem Kolff, invented this machine. His goal was to find a way to save the lives of people whose kidneys had failed. He kept improving his invention with the cooperation of other doctors and by 1945 he began treating his first patient, a 67-year-old woman whose kidneys had failed. She lived another seven years with the new treatment. Of course the dialysis machine has been greatly improved since those early years. Basically, a tube is inserted into an artery, so blood passes through the dialysis machine, then the blood returns through another tube into a vein and back into the blood stream. The machine's main function is to filter impurities from the blood. A person on dialysis normally experiences the process about three times each week. This person is either waiting for a kidney transplant or will be on dialysis for rest of his/her life.

Chapter 7
SENSE OF HEARING

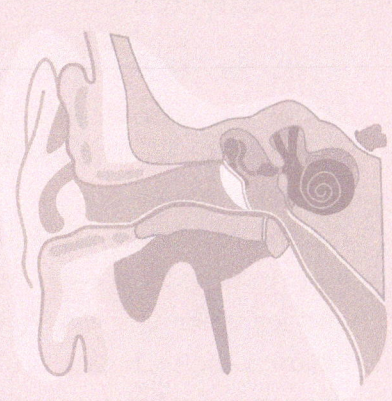

Our sense of hearing is an amazingly complicated process, designed to do several things at once. For instance - the entire construction of the ear (the external ear, the middle ear and the inner ear) would be of no value if it were not connected to the brain. The brain then interprets the sound waves, which vibrate our eardrum, then the brain lets us know what we have heard. But before those vibrations reach the brain an incredible series of events takes place. The following drawing of the entire ear will help us understand our sense of hearing (figure 1).

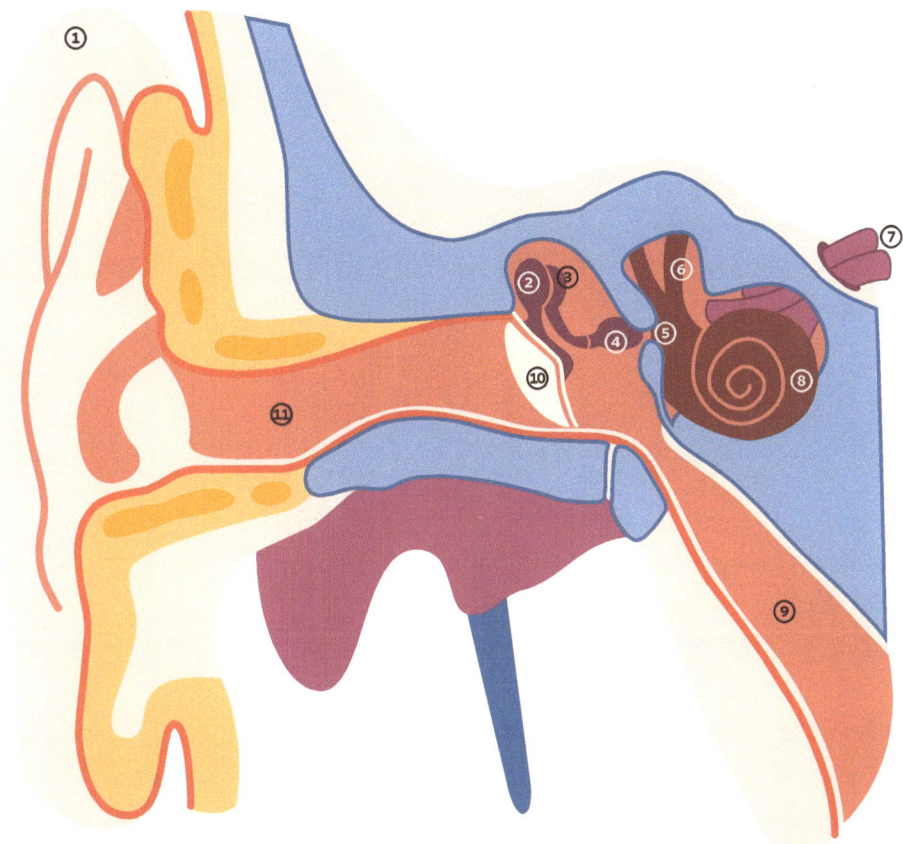

Chapter 7, Figure 1

1. External ear
2. Malleus
3. Incus
4. Stapes (numbers 2, 3 and 4 are known as the ossicles)
5. Oval window
6. Semicircular canals - related to balance
7. Cochlear nerve
8. Cochlea - vibration changes to electrical signals here
9. Eustachian tube - Connects middle ear with throat - equalizes air pressure between ear and throat.
10. Tympanic membrane (eardrum)
11. Auditory canal

Figure one shows the external ear and the passage to the eardrum. The eardrum is also called the tympanic membrane. The middle ear begins at the eardrum and ends at the oval window of the inner ear. The inner ear begins at the oval window and contains the cochlea and the organs of balance. The cochlea is involved with hearing, and the organs of balance contain three semicircular canals and other parts.

Looking at the external part of the ear we notice that the earlobe is designed to catch sound waves and funnel those waves to the middle ear. The sound waves pass through the auditory canal to the eardrum. This canal passes through an opening in the bony structure on the side of our head.

The middle ear contains three tiny bones called ossicles. These bones are specially designed to work together to transmit sound waves to the inner ear. The first bone, the malleus, is attached to the eardrum - the tympanic membrane - and also vibrates against the middle bone, the incus, and in turn, the incus vibrates against the stapes, the last of the three bones. The stapes sends the sound signals through the oval window to the part of the inner ear called the cochlea, which is shaped somewhat like a snail. The inner ear is surrounded by a bony structure, which protects the sensitive parts of the inner ear.

So far sound waves have passed through the three parts of the ear but have not yet been interpreted. If this were the end of the process we would not understand what the sound waves mean. The sound waves must move from the inner ear to the brain through the cochlear nerve. But the cochlear nerve receives information from that snail-like part, the cochlea, through a complicated process. The cochlea contains fluid-filled cavities and the central cavity contains tiny hair cells, which convert sound waves into electrical nerve impulses. The cochlear nerve transmits these impulses to the brain. But those electrical impulses don't tell us what we have heard; the brain now must interpret the message it has received.

You will notice in figure one that the cochlear nerve joins with the vestibular nerve before these two nerves reach the brainstem. We'll learn more about the vestibular nerve when we look at how the ear also sends messages to the brainstem where balance is maintained. Those two nerves pass through an opening in the bone and make contact with the brainstem. There are many important medical terms/names that doctors use when describing how the nerve impulses reach the part of the brain which interprets sound. We can by-pass those terms because we only want those impulses to reach their destination in the brain, the auditory cortex, and the auditory association cortex, also known as the acoustic imaging center (figure 2).

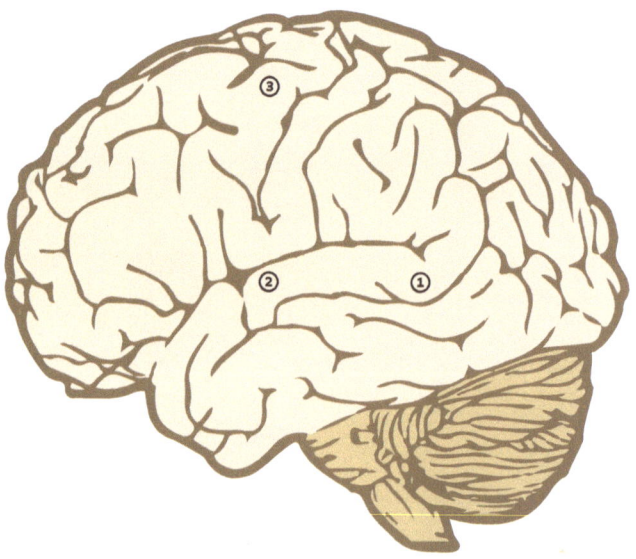

Chapter 7, Figure 2

1. Auditory association cortex – The interpretation of the meaning and significance of sounds is dealt with here.
2. Primary auditory cortex - Features of sound like pitch and rhythm are processed here.
3. Right cerebral hemisphere

This auditory center analyzes both high and low frequency sounds, interpreting in a split second what we have heard.

So we have moved sound waves from the eardrum, through the three bones in the middle ear, to the cochlea in the inner ear, through the cochlear nerve to the brainstem, and up to the auditory cortex where sound is analyzed and interpreted. And all of this takes place in a split second. For instance, if someone is standing a few feet from you and says "hi", you know immediately what that person said although that sound traveled the route we have just described. And there is something else worth considering; the sound is interpreted in our brain but we hear the sound at its place of origin. Our sense of hearing is a magnificently designed process and the more we learn about it, the more amazed we are.

Now we can go to figure three and learn about our body's balance mechanism. The three semicircular canals are hollow bony canals, arranged at approximate right angles to each other, and they finally join with the vestibular nerve. Through an obviously complicated process this nerve has received signals related to movement of the head and sends these signals to the vestibular nuclei, part of the brainstem. The brainstem lies at the junction of the spinal cord and the brain. A person's balance is controlled in the brainstem through signals it received from the inner ear by way of the vestibular nerve.

We can also recognize the fact that our sense of hearing is designed so that we have the reception part on both sides of our head and the signals from both inner ears travel to the brain for interpretation. It is also important to realize that openings through bone appear at identical places on both sides of the skull. The outer ear, middle ear, inner ear, small openings through the skull and connection with the brain are specially designed to give us our much-needed sense of hearing.

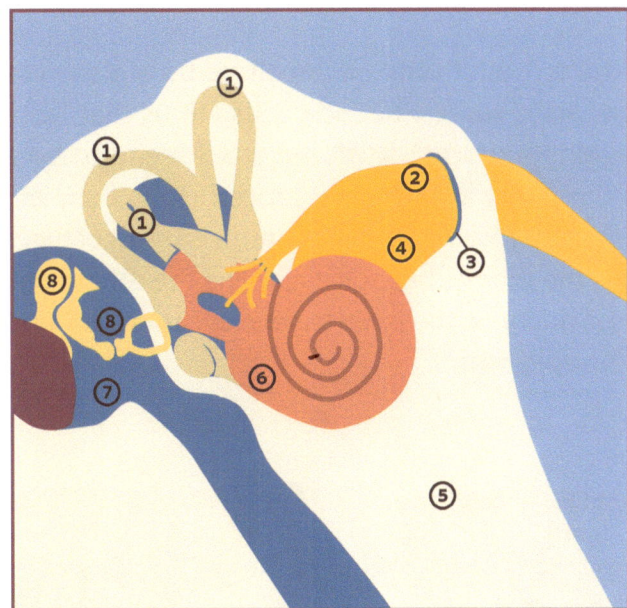

Chapter 7, Figure 3

1. Three semicircular canals
2. Vestibular nerve
3. Opening in bone through which nerves pass to brainstem
4. Cochlear nerve
5. Bone
6. Cochlea
7. Chamber for middle ear
8. Ossicles

Chapter 8
SENSES OF TASTE AND SMELL

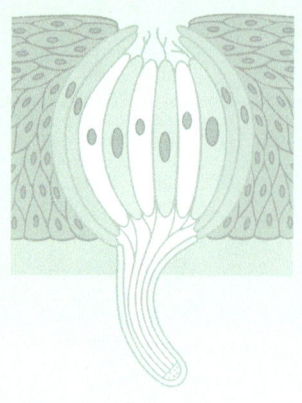

The senses of taste and smell are usually seen as separate senses. But scientists tell us that taste is eighty percent smell, but smell seems to be a completely independent process. Taste buds are obviously located in the mouth, with most of them on the tongue but some also line the soft palate, the pharynx, the inner surface of the cheeks and the epiglottis.

In the mouth the cells that make up the taste buds are called gustatory cells, (figure 1) but in the nasal area there is a small patch of cells in the roof of the nasal cavity. This patch of cells, about one square inch in size, is called the olfactory epithelium (figure 2). It contains between 10 million and 100 million olfactory receptor cells. In both of the senses, chemicals cause the receptor cells to send messages to the brain for interpretation.

Chapter 8, Figure 1

Typical gustatory cell on the tongue
1. Pore
2. Gustatory hair - Microscopic hairs that are bathed in saliva-containing food molecules. They send their signals through the gustatory cells to the nerve fibers and then to the brain.
3. Epithelial cells make up the outer layer of the tongue.
4. Nerve fibers send signals to a region of the brain called the thalamus.

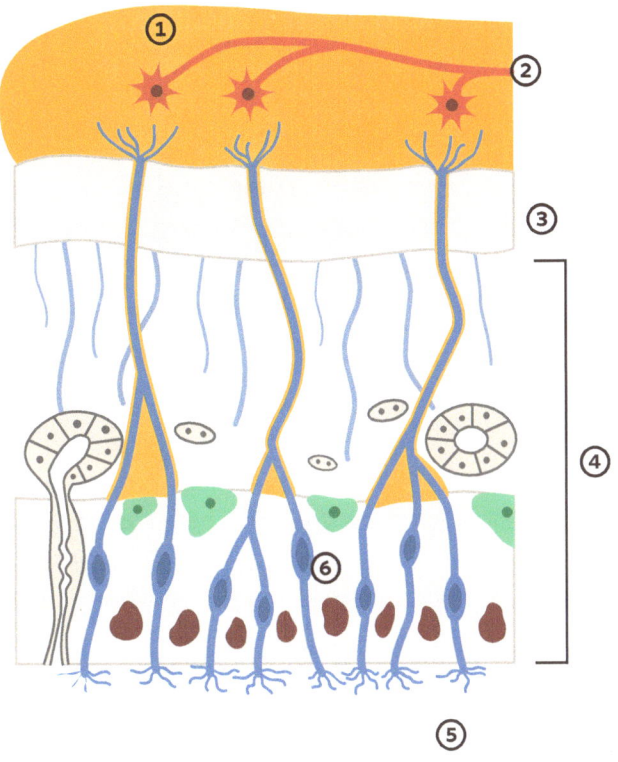

axons (nerves) to the olfactory bulbs (figure 3). The bulbs are located just above the nasal cavity and the signals sent to the bulbs from the receptor cells have traveled through small openings in the ethmoid bone between the nasal cavity and the olfactory bulbs. From the bulbs the signals travel to the thalamus and then to the olfactory cortex of the brain. It is then that we know what we have smelled, and this transfer of information has taken place in microseconds.

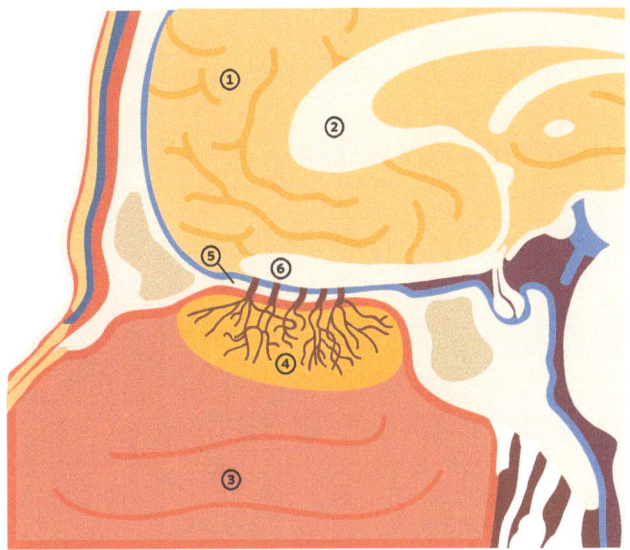

Chapter 8, Figure 2

1. Two olfactory bulbs process information from the olfactory nerve fibers and send that information to the brain, to the olfactory cortex, where it is interpreted.
2. Nerve fibers of olfactory tract send signals to the brain.
3. Ethmoid bone - Separates olfactory epithelium from the olfactory bulbs. Tiny, almost microscopic openings in the bone allow the olfactory nerve fibers to send their signals to the olfactory bulbs.
4. Olfactory epithelium contains millions of olfactory receptor cells. It is located at the roof of the nasal cavity.
5. Nasal cavity
6. Olfactory receptor cell senses odor molecules.

Chapter 8, Figure 3

A closer look at the olfactory bulbs and nerve fibers:
1. Frontal lobe of the cerebral hemisphere
2. Thalamus
3. Nasal lining
4. Olfactory nerve fibers
5. Ethmoid bone lines top of nasal cavity.
6. Olfactory bulbs - Signals travel from these bulbs first to the thalamus and then to olfactory cortex.

We will look at the sense of smell first. We mentioned the olfactory epithelium in the roof of the nasal cavity. Some of the molecules in the air that we breathe become attached to the olfactory epithelium. It has a mucous covering. Microscopic hairs on the cells send signals through

Before we look at the sense of taste we should know that molecules from the food that enters the mouth will enter the nasal cavity through the nose and through the opening at the soft palate at the back of the mouth. The soft palate closes as we swallow, but it is open as we chew.

As food is chewed chemicals in the food molecules are momentarily caught by the taste buds in saliva in the mouth. In each taste bud are sensory receptor cells called gustatory cells (figure 1). A tiny hair like projection extends out of each taste bud. These projections send a signal to sensory dendrites at their bases. From there the signal is passed to the brain stem, then to the thalamus then to the gustatory cortex in the brain.

These signals plus the signals from the sense of smell tell us what we taste. And again we are impressed by the way the brain is involved in all of the body processes. The design that gives us the sense of taste and smell seems to have a plan behind it, creating first the need for the senses and then the complicated arrangement that completes the task of answering that need.

Here is another interesting point, and it shows how dependent taste is on smell. When a person with a bad cold complains that food doesn't taste the way it should, it is because most of the sense of smell has been overpowered by the cold, and that little one square inch patch called the olfactory epithelium cannot function as it is supposed to. But when the cold goes away the sense of taste works better because the sense of smell is "back on board" and food tastes like we expect it to.

Chapter 9
SENSE OF VISION

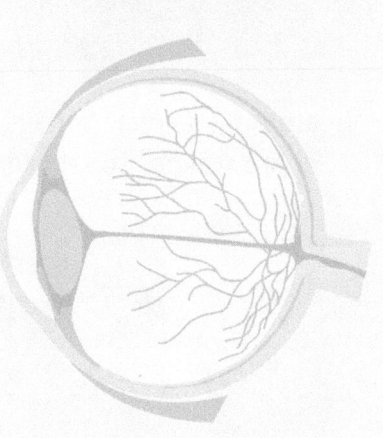

Looking at all of our senses, vision is the one which we rely on the most to identify our surroundings, and realize how other people react to us. We recognize our friends and watch events unfold. With the aid of a mirror we see who we are. Our other senses such as hearing, taste and smell may aid us in realizing what our surroundings are like, but vision probably plays the most important role. More than any of our other senses, vision helps us define who we are.

In order for our eyes to function as they are designed to, all of their "parts" must work together flawlessly. But that is only part of the story. The back part of our brain cooperates with each eye to interpret what we are seeing and immediately tells us what is in our field of vision.

During an eye exam the doctor looks into each eye and sees much more than we might suspect. However, we can get a pretty good grasp of how our eyes are designed, how that design allows them to do what they are intended to do and how the brain enters into the process.

The eyeball is a sphere, about one inch in diameter that sets inside the opening in the front of our skull known as the eye socket (figure 1). Another opening right behind the eyeball allows the optic nerve to leave the back of the eye and send signals to the back of our brain, completing the cycle that gives us the sense of sight or vision. And it is important to note that an eye may be perfectly formed but without that connection with the brain, it cannot function. We would have no vision.

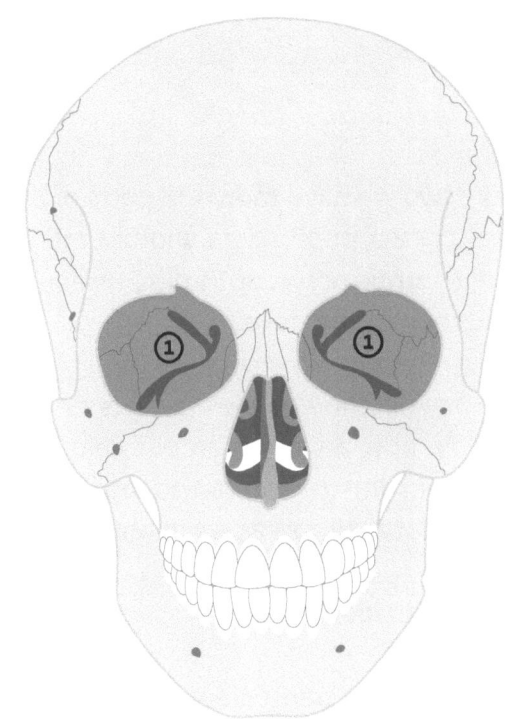

Chapter 9, Figure 1

Frontal view of the skull
1. Eye Sockets - The eyeballs are set in each socket.

Figure two gives us a pretty good idea about the design of the eye. As it sets in the socket only a small part is actually in the open. When you look into another person's eyes you see the white, the color of the eye and the pupil. These three parts

35

are only a small percentage (about 16%) of the entire sphere. So let's look inside this object we call the eyeball for a better understanding of what its composition is and how it works.

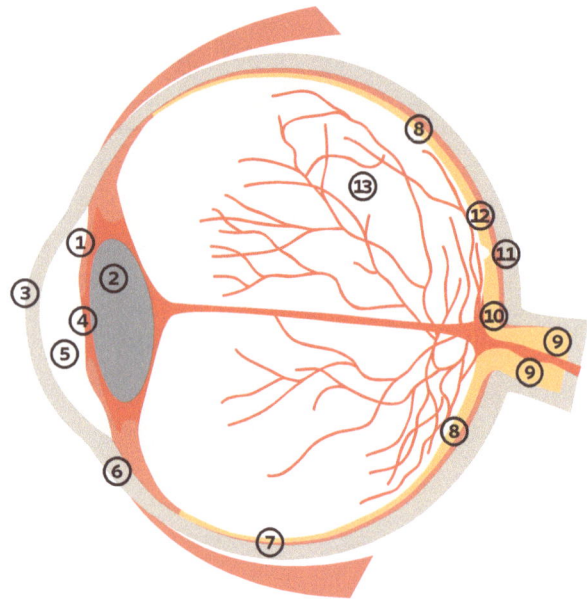

Chapter 9, Figure 2

1. Iris
2. Lens
3. Cornea
4. Pupil
5. Aqueous humor
6. Sclera
7. Choroid
8. Retina
9. Optic nerve
10. Optic disc
11. Fovea
12. Rods and cones in retina
13. Vitreous body

In figure two we have shown 13 parts of the eye: iris, lens, cornea, pupil, sclera, choroid, retina, rods and cones, optic nerve, optic disc, vitreous body, aqueous humor, and fovea.

We will first describe/define the parts mentioned, then show how an object in our field of vision becomes an image we can identify, and this process is not as simple as we might think.

But first - definitions:

The cornea is the transparent layer covering the front of the eye.

The iris is the colored part of the eye immediately behind the cornea.

The pupil is actually an opening in the iris that allows light to enter the eye. The pupil is wide open in dim light and smaller in bright light.

The lens is a transparent structure that focuses images onto the retina.

The sclera is the outer layer of the eyeball.

The choroid is the middle lining of the eyeball.

The retina is the inner lining of the eyeball. The retina contains rods and cones, which respond to light that has entered the eye.

Rods are microscopic projections in the retina that respond to images in the dark and send their response to the optic nerve. Rods do not respond to color.

Cones are microscopic projections that send color responses to the optic nerve. Rods are sensitive to diminished light, which causes objects to lose their color at night. Cones respond to bright light so the brain interprets color images from the cones.

Rods and cones are called photoreceptors and there are 125 million in each eye.

The optic nerve originates in the back of the eyeball and sends signals from the retina to the brain.

The optic disc is the point at which the optic nerve connects with the retina.

The vitreous body is the largest expanse of the eyeball. It is filled with a clear jelly-like fluid called vitreous humor.

Aqueous humor is the fluid in front of the lens.

Fovea is a slight depression in the retina where maximum light is focused.

The eyes and brain tell us what we see. An image in the form of light enters our eyes and is directed to the retina. That light has first passed through the cornea and the aqueous humor. It then passes through the lens where it is further defined before it is directed to the retina. When it reaches the retina it is well focused, but it is upside down and the sides are also reversed. Rods and cones in the retina have received the image and pass that "information" to the optic nerve. The optic nerve transmits that light/image to the back part of the brain where it is interpreted and we now know what is in our field of vision.

The brain has received light from the optic nerve and has transformed that light into a picture. What entered the eye was light and not until that light reached the brain was it made into a picture, turned right side up and the reversed sides were corrected.

Some other processes take place that are worth noting. We learned that the lens further defined the light entering the eye. But if we are looking at an object at a distance, the lens is flattened by certain muscles and ligaments. If we are looking at an object up close, the lens becomes rounded, shaped almost like a ball.

Try a little experiment. Look at an object a good distance from you. Now, look at something you are holding in your hand. Your eyes immediately focused on each object.

If you look at something 300 feet away, then something 30 feet in the distance, and then something in your hand, you now realize that the lens was being controlled by muscles and ligaments producing necessary changes in the shape of the lens although we are ordinarily unaware of those changes.

The lens is a remarkable part of the eye, but we should note here that regularly scheduled eye exams can alert us to a condition that affects most elderly people - cataracts. Cataracts affect the lens by gradually showing a cloudy covering, which is a degeneration of the lens structure caused by calcium crystals. Now the lens can actually be replaced by an artificial lens, which almost always gives the patient much better vision.

There is another feature we probably don't know about, but it is at least interesting when we understand what is happening. We learned earlier that the optic nerve sends light images to the back of the brain. However, the back of the brain in both halves receives this information. The left half of each eye sends information to the right part of the brain and the right half of each eye sends information to the left half of the brain through the optic nerve. The point at which these optic nerves cross is called the optic chiasma (figure 3).

Another feature worth noting - blood vessels enter each eye through the opening at the back of the eye where the optic nerve makes its exit. The eyes need blood just as all of the body needs this vital fluid.

We go through life using our sense of vision every day but have probably never taken the time to try to understand just how amazing and complicated it is. All of these parts/pieces work together to give us a sense we may too easily take for granted.

Intelligent Design or Non-Intelligent Design?

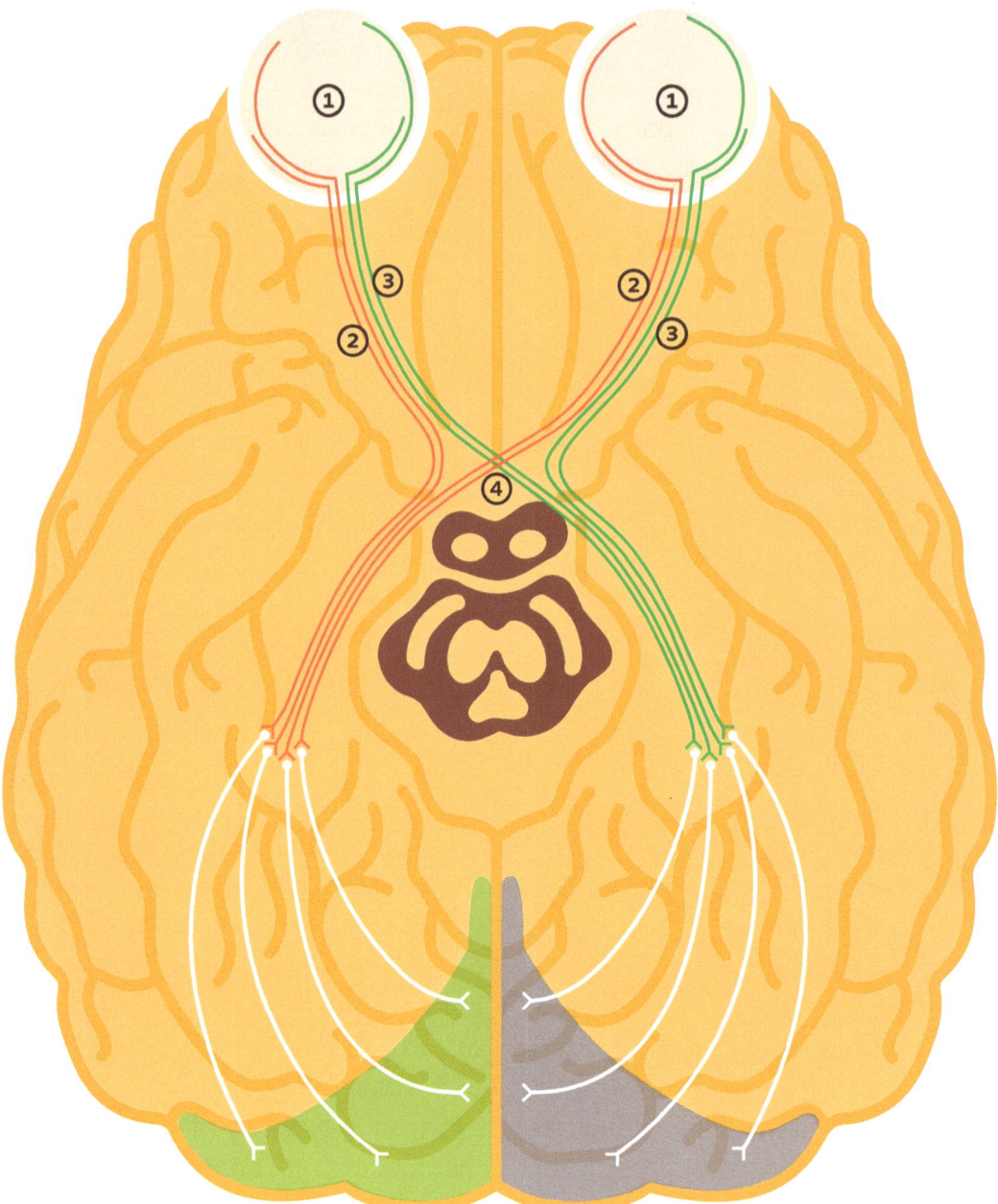

Chapter 9, Figure 3

This drawing shows a cutaway of the optic nerves and how they send information to the back part of the brain where everything is interpreted. The left half of each eye sends its signals to the right part of the back of the brain and the right half of each eye sends its signals to the left part of the brain. The point at which the optic nerves cross is called the optic chiasma. The signal that reaches the brain is inverted all the way around but the brain turns it "right-side-up" as well as interpreting what is in our field of vision. And this interpretation is immediate. This drawing is a view from below.

1. Eyeballs
2. Optic nerves taking signals from the right half of each eye.
3. Optic nerves taking signals from the left half of each eye.
4. Optic chiasma

Chapter 10
SKELETON, BONES AND TEETH

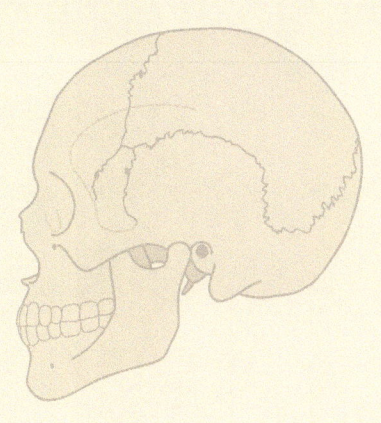

The human skeleton is a structure of 206 bones (figure 1) and scientists/doctors divide the skeleton into two main parts - the axial and appendicular skeletons. The axial skeleton includes the skull, backbone or spine and the rib cage. The appendicular skeleton includes the arms and hands, the legs and feet, and the pelvic and pectoral girdles. These two girdles attach the arms and legs to the axial skeleton. You can easily identify the skull, rib cage and spine in figure one, as well as the arms and legs, pelvic and pectoral girdles.

The skull protects the brain and hearing, the spine protects the delicate spinal column and the ribcage protects the heart and lungs.

There are different kinds of joints that hold the skeleton together, and each of these joints allows a certain kind of movement. But there is one exception. The cranial part of the skull is composed of eight bones, joined together, but these bones do not move (figure 2). These solid joints are actually called sutures, and are cemented together by thin bands of fibrous tissue preventing any possible movement.

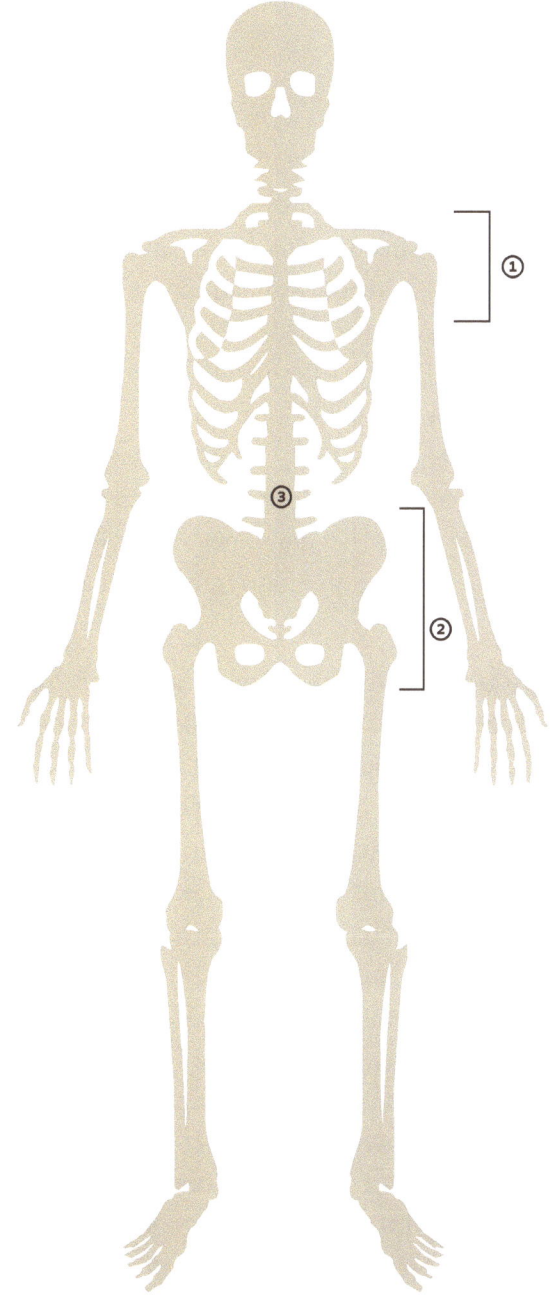

Chapter 10, Figure 1

Frontal view of the human skeleton
1. Pectoral girdle
2. Pelvic girdle
3. Spine

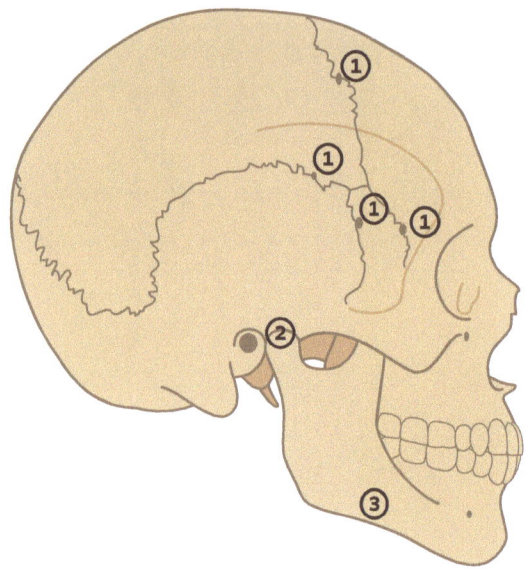

from the compact bone but the bone marrow actually helps to keep the human body alive. The marrow produces red blood cells, white blood cells and platelets. The red marrow turns out about one hundred billion new blood cells every day. The red cells transport oxygen from the lungs to the entire body. They also transport carbon dioxide from all parts of the body to the lungs where it is eliminated through the act of breathing. White blood cells fight disease and platelets help in the blood clotting process.

Chapter 10, Figure 2

Joints of the skull
1. Sutures - This drawing of the skull shows some of the bones that are joined together by sutures, fibrous tissue that bonds the skull bones together.
2. Mandibular joint - The mandibular joint shows the only bone in the skull that moves.
3. Mandible (jaw bone) - This bone moves up and down, sideways and in and out. When we chew our food we know that our upper and lower teeth are involved, but only our lower teeth actually move.

Before we look at the different kinds of joints in the skeleton, we'll examine the bones themselves. Bones grow because cartilage at the end of a bone grows longer, but the part of that cartilage attached to the bone actually turns into bone, gradually adding to the length of that bone. Newborn babies have arms and legs just a few inches long, but they begin growing as described. This growth continues until a person reaches age 20 – 25.

Just a brief look at the composition of a bone shows a hard outer layer called compact bone and a soft inner part containing bone marrow (figure 3). A thin membrane called the periosteum covers the outside of the bone. A bone's strength comes

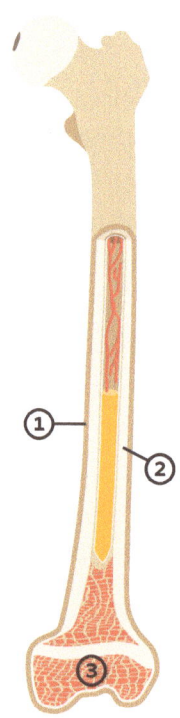

Chapter 10, Figure 3

This drawing shows a cutaway view of a typical bone in the human body.
1. Periosteum - a thin covering on the outside of a bone.
2. Compact bone - the hardest material in the human body except for the enamel covering our teeth.
3. Bone marrow

Bones are described as living organs. Along with the work of bone marrow, there are two kinds of cells at work in/on every bone - osteoblasts and osteoclasts. Osteoblasts build new bone and osteoclasts absorb old bone cells and release their calcium into the blood stream where it will be used to build new bones. The material not used will be removed from the body as waste. It should be noted here that the compact bone at the outer surface is the hardest substance in the body except for tooth enamel. We also note that only about fourteen percent of body weight is composed of bone. If the entire skeleton were solid bone it would be too heavy to perform as it does.

Examining the design of the different kinds of joints lets us understand what takes place when we throw a ball, when we run or when we chew our food.

- Bend your arm or leg, move your fingers or toes and you have moved a hinge joint.

- Hips and shoulders have ball and socket joints, which allow for a wide range of motion.

- A planar joint is one in which there is limited motion such as the joint holding the breastbone to the collarbone. The scientific term for the breastbone is the sternum and the collarbone is called the clavicle.

- The joint which allows the neck's wide range of motion is called a pivot joint.

- An ellipsoidal joint allows limited movement such as the wrist joint.

- The joint connecting the thumb to the wrist is called a saddle joint, allowing the thumb to move backward and forward and also to rotate.

When two bones in a joint move against each other they are separated by a lubrication, which allows for easy movement. Bones are held together by ligaments, which are strong fibrous tissues. Tendons are strong tissues that secure muscle to bone and cartilage is the material at the end of a bone which, among other things, is covered by the gel-like lubricating substance, mentioned above, which reduces friction when joints move.

When muscles stretching from one bone to another contract, they cause the joint to move. This motion happens because the brain sent a signal to that muscle to contract, and the brain sent that signal because another part of the brain decided it wanted that joint to move. This brain-muscle interaction is discussed in detail in chapter three.

The skull is the most complicated bone structure in the human body. We might have a tendency to think of the skull as one globe-like bone. However, there are eight bones that make up the part that protects the brain and hearing, and fourteen bones make up the face and jaw; the jaw is the only bone that is freely movable. This bone moves from side-to-side, back and forth and up and down.

The frontal view of the skull shows how complicated this mass of bones is. You can see the sutures that fasten the cranial bones together. In chapter nine, vision, we discussed the sockets in which the eyes are located. These sockets are surrounded by more than one bone, all coming together to form the opening for the eyeball, part of our sense of vision. (We also learned in chapter nine that optic nerves which originate in the eyeball must communicate with the back part of the brain in order to complete the sense of sight or vision.)

The spine or backbone, is another part of the skeleton that deserves our attention, but its relationship to the Central Nervous System is covered in chapter three. However, as part of the skeleton we must recognize its importance as a supporting structure. The spine serves as the support chain that extends from the pelvic girdle up to the head. The spine is divided into four main parts (figure 4). The neck is supported by the six cervical vertebra. The next twelve vertebra, thoracic vertebra, form the outward curve of the back. This is also where the ribs connect to the backbone. The thoracic vertebra are stronger than the cervical vertebra, but the strongest division is the five lumbar vertebra. These vertebra support most of the body's weight. The lowest part of the spine is the pelvic curve, containing five fused bones called the sacrum and four fused bones called the coccyx, or tailbone.

Muscles are attached to the spine to hold it upright. There are 26 bones, vertebra, in the three regions named above, not including the fused bones in the pelvic curve. And as noted in chapter three, the spinal cord is enclosed in the spinal column where it is protected from outside forces.

The human skeleton is designed to give the body movement as needed, to give support as necessary and to protect vital organs from damage caused by external pressure.

Before we end this chapter, let's look at the way our teeth are formed and their role in keeping the human body functioning and healthy.

32 teeth form the upper and lower arrangement in the human mouth. All of these teeth are attached inside the jawbones by roots and the teeth are surrounded by soft tissue known as gum. (As mentioned earlier, the hardest substance in the human body is the tooth enamel.) Of the 32 teeth, four are known as wisdom teeth, which are the last teeth in the upper and lower gums. These teeth are unique in a special way - sometimes they never work their way through the gum, sometimes they come through the gum and cause no trouble and

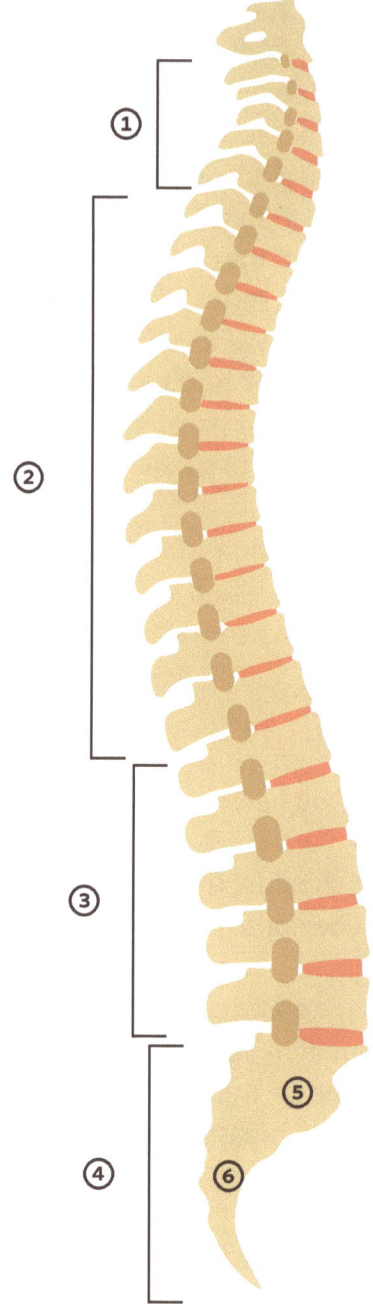

Chapter 10, Figure 4

The spine is divided into four main parts.
1. Six cervical vertebra
2. Twelve thoracic vertebra
3. Five lumbar vertebra - They support most of the body's weight.
4. Pelvic curve - Sacrum and coccyx are found here.
5. Sacrum - five fused bones
6. Coccyx - four fused bones. The coccyx is also called the tailbone.

sometimes they become problems while still not visible and at other times, even though they are through the gum they become painful and must be extracted.

We are all familiar with trips to the dentist for many different reasons, from having our teeth cleaned, having cavities filled, oral surgery including extractions, and for younger children - getting braces to properly align all of the teeth - both upper and lower.

There is also the special expectation that comes to five, six or seven year olds - losing their baby teeth. There are ten upper and ten lower baby teeth, called primary teeth. They are replaced by adult teeth almost immediately, in fact most of the time when baby teeth "fall out", their replacement is already visible.

Two other facts are noteworthy, teeth are living parts of our bodies with blood vessels entering all of them and servicing them as long as they remain in the mouth. Teeth also hold nerves (figure 5), which tell us something is wrong by giving us a tooth ache which means we don't waste any time making a trip to the dentist.

Finally, take another good look at the skeleton. It is a remarkable assemblage of bone, muscle, cartilage and ligaments - all working together to present the human body as an amazing, complicated, specially designed machine.

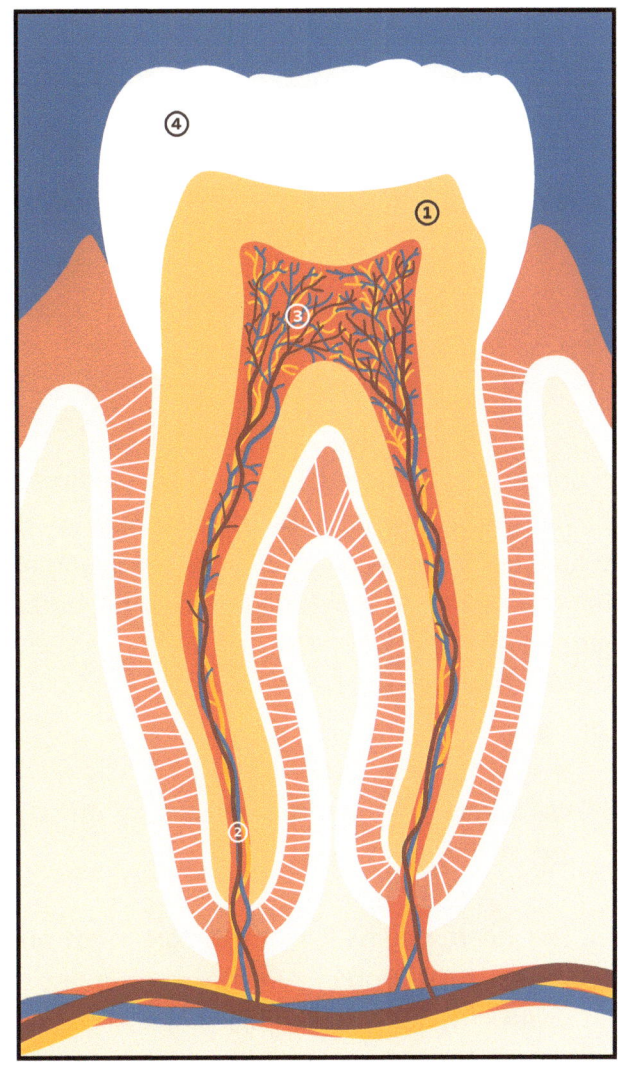

Chapter 10, Figure 5

Anatomy of a tooth
1. Dentin
2. Root canal - When a person has an abscessed tooth (infected tooth), a dentist can usually save the tooth by performing a root canal by drilling out the root canal and filling it with a substance that hardens and will not decay.
3. Pulp chamber - Blood vessels and nerves are located here.
4. Enamel - Hardest material in the human body. Bone is next to the hardest.

Chapter 11

MUSCLES

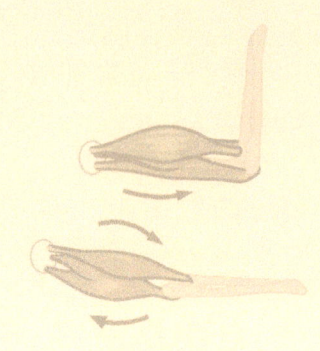

In order to gain some kind of understanding of the human body's muscular architecture, we need to look first at a few facts about these structures that make up more than fifty percent of the body's mass, or weight.

For instance:

- We breathe day and night because muscles are constantly at work.

- The heart begins beating before birth and never stops working until the moment of death - the heart is composed almost entirely of muscle.

- The food we chew and swallow moves down the esophagus because of muscular action.

- Our eyes respond to light and darkness and they focus on objects at different distances because of muscles.

- Muscles that cover the outer surface of the human body serve as protection for all of the internal organs along with the skeleton.

- Every time we change facial expression muscles are at work.

- Muscles cause lymph fluid to move from the outer reaches of the body finally to the heart.

- Muscles move food in the stomach toward the small intestines, from the small intestines to the large intestines and finally out of the body. This action is called peristalsis.

- Muscles in arteries are involved in controlling blood pressure.

The above list gives us an idea just how important the muscular makeup of the human body really is. However, in reading that list we are left wondering about the types of muscles in every body.

There are three types of muscles in the human body. Smooth muscles line every organ in the body, and they also cover arteries and veins. Skeletal muscles are attached to bones and cause the joints to move receiving nerve signals from the brain. Cardiac muscle is a special kind of muscle that is found only in the heart. All three types of muscle are made up of long, thin cells, which also can be called fibers.

Muscles are also described as voluntary and involuntary. Voluntary muscles work because a human being intends for them to accomplish certain tasks. For instance, a person decides to lift his arm (figure 1). The action is instantaneous but the following steps are required. The decision to move the arm is recognized in one part of the brain, it is relayed to another part of the brain, then down through nerves in the spinal cord and out to nerves in the muscle cells in the arm. The arm will then move as the person wanted it to. Chapter three covered just how involved the brain/central nervous system is in all that the body accomplishes.

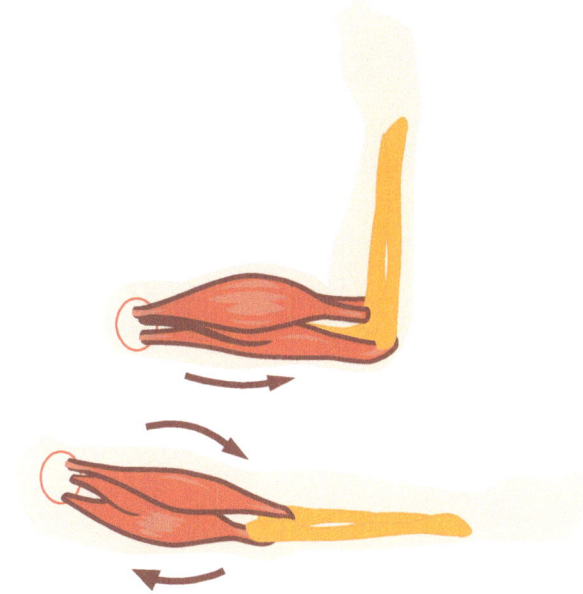

Chapter 11, Figure 1

Flexed and extended muscles
The arm is shown flexed (top view) and extended. Either movement required the person to decide to move his/her arm. Then the brain sent a signal to the muscles either to contract or extend, but only after the person decided to move the arm.
Every time a person decides to voluntarily move any part of his/her body, the brain is involved.

Involuntary muscle movement occurs when action is required but the person doesn't decide to cause the movement. For instance - the heartbeat and peristalsis in the esophagus.

Involuntary muscles work differently. When a person swallows a bite of food (called a bolus; see chapter five), the process called peristalsis begins its intended work. Smooth muscles line the esophagus and examining their structure shows how amazing the design is. We know from chapter five that the esophagus is simply a tube that connects the mouth with the stomach. The esophagus is lined with smooth muscle that runs the entire length. But another "collection" of smooth muscle encircles the esophagus. As the bolus enters the esophagus, the muscles in front of the food relax making room for the food while the muscles behind the food contract forcing the food along. This process continues until the food reaches the stomach. If that person is eating a meal, the esophagus continues its work. The only decision the consumer of the meal made was to swallow, the esophagus took over from there, so they are involuntary muscles.

The heart is another example of involuntary muscle action. The heart beats constantly, sixty to seventy times per minute and we don't have anything to do with the contraction of the heart muscle. It is designed to work quite efficiently. However the brain is involved - if a person is exercising and his muscles are needing more oxygen, the brain makes the heart beat faster to collect more oxygen from the lungs and pump it to the part of the body needing that oxygenated blood. The brain had received a signal from the muscles involved that more "fuel" was needed. Blood brings glucose (sugar) to muscle cells where oxygen breaks it down into energy. But when demand for energy is high the muscle cells can break down glycogen, another kind of sugar actually stored in the muscle.

When physical exercise demands more than the muscles can deliver, muscular fatigue sets in. The brain can still send a message for the muscles to perform, but the muscles cannot do what the brain expects. In this case, the heart is not beating fast enough to send the needed oxygen, and the muscles cannot respond to the brain's message. This fatigue is actually caused by a build-up of lactic acid, which slows down muscular action.

At/in every muscle cell arteries, veins, capillaries and nerves come together to receive messages from the brain, to supply the cell with energy by bringing oxygen to the cell and to remove carbon dioxide. (Capillaries are covered in chapter two.)

Finally, we must recognize the importance of exercise. Exercise does not build new kinds of muscle but it does increase the size of existing muscle. Weight lifters have the same muscles as everyone else but their muscles are in an expanded state because of their desire to "build" muscle. But a person doesn't need to exercise/lift weight to the extent just described. It is simply important to be concerned about muscle tone, keeping muscles "fit" to maintain a healthy body. There are some people who seldom exercise, sometimes because they simply don't want to, and sometimes because they have jobs that require sitting for long hours in offices that don't allow for much movement. If these folks don't get into some kind of exercise regimen their muscles will eventually begin to atrophy, they shrink in size and the body begins to show it is not getting the right amount of exercise.

Although this chapter presents just a brief discussion on muscles, we do begin to understand why muscles are so important and why we need to take good care of our bodies.

A final word - - there are plenty of sources to learn about exercise programs available to all age groups, and any doctor will help move us in the right direction.

Chapter 12
IMMUNITY AND LYMPHATICS

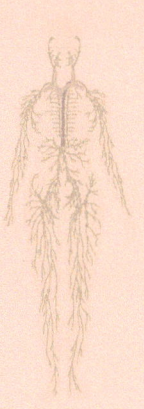

Trying to gain an understanding of the human body's immune system could be described as overwhelming/almost impossible. In fact, you may want to read this chapter three or four times because of the many different tasks performed by the immune system.

When an unwelcome foreign substance enters our body, it can be attacked in many different ways by many different agents. And some amazing changes can take place in the defense system as it attacks the foreign invader/s. White blood cells are the main defenders - at least that is their main role and it is interesting to realize the "communication" that takes place while the body is attacking these uninvited "guests".

Before we begin looking at white blood cells we should examine the lymphatic system. And this will come as a mild surprise to many people - the lymphatic system is a one-way conducting system in the human body (figure 1). Its vessels follow the same route as arteries and veins for the most part, but the fluid - called lymphatic fluid - in these vessels is not pumped, it is moved by muscular action and the fluid is kept from moving backward by one-way valves. Lymphatic fluid, which surrounds cells, is found throughout the body. It is found in the blood and some leaks from capillaries and is picked up in the lymphatic vessels and travels to lymph nodes where it is filtered.

Chapter 12, Figure 1

The lymphatic system
The lymph vessels follow the same general direction as blood vessels but only travel from all parts of the body to the heart. They are inbound only. In this drawing the lymph nodes are the slightly enlarged pieces on the lymph vessels and are distributed throughout the lymphatic system. They filter lymph fluid for impurities before the fluid reenters the blood stream. White blood cells are embedded in the lymph nodes and help to destroy the impurities.

In the lymph nodes the fluid will give up foreign particles that the blood picked up as it traveled through the body. We note that lymph fluid only travels in a direction that takes it to the heart; it does not travel out from the heart as a separate liquid as blood does, so we can call it a one-way conducting system.

cells are manufactured. These blood cells move from the marrow to the circulatory system and immediately begin their work. (White blood cells are also known as corpuscles or leucocytes, and as important as they are, they only make up about one percent of the blood supply, while red cells make up at least forty percent of blood volume.)

Almost all blood cells are manufactured in the bone marrow and all come from a single type of cell called a stem cell. The spleen filters blood to trap damaged red cells but it also produces lymphatic cells and antibodies which help fight infections.

Back to white blood cells: once they are made, they basically go to one of three places - the blood stream, the spleen or the thymus (in the latter two cases they actually will change into other kinds of cells.) When a foreign substance gets past the body's first line of defense - the skin and mucous where many "foreigners" are trapped - this invader will now encounter the body's second line of defense. These invaders are also called pathogens and antigens. The body's first and second line of defense is known as the innate defense system.

It would be easier to understand the role of white blood cells if they simply remained in their original form when they were first produced, but this is not the case. The following terms apply to white cells as they play out their role in combating disease:

B Cell, T Cell, Basophil, Lymphocyte, Macrophage, Natural Killer Cell (called an NK cell) Neutrophil and Phagocyte. (As you encounter these different names in this chapter remember that they all originate as white blood cells).

When an invader (pathogen), has moved past the skin and mucosal lining, the internal innate defense system takes over. Phagocytes and natural killer cells will now be involved. We will look at the

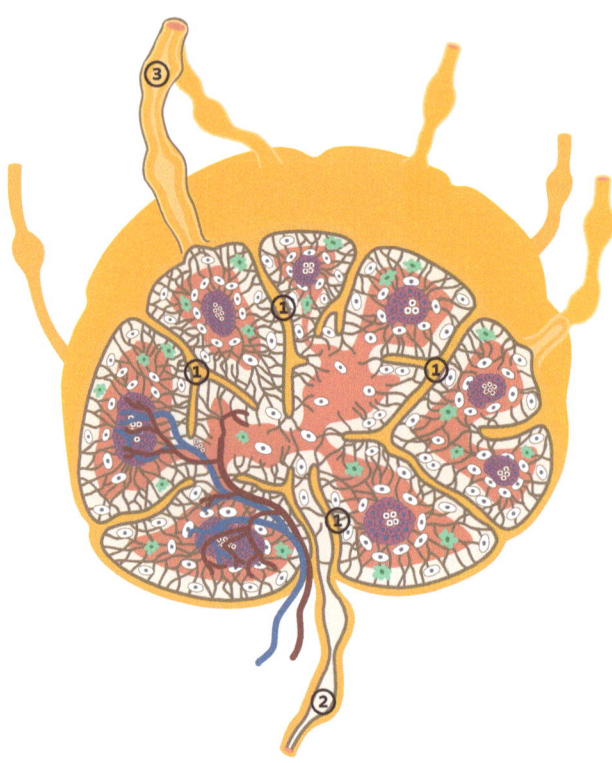

Chapter 12, Figure 2

A typical lymph node
1. Trabeculae are fibrous tissues that divide the lymph node into segments. Inside these segments are various cells that fight disease-carrying invaders. When the fluid leaves the lymph node it has been cleansed of invading organisms and returns to the blood stream.
2. Efferent lymph vessel transports filtered fluid from the node.
3. Afferent lymph vessel transports lymph fluid to the lymph node.

(Figures 3 and 4 are drawings of photos taken from a microscope showing how the body fights disease.)

The main "players" in the body's immune system are the lymphatic system, the spleen, the thymus, the circulatory system and bone marrow. Since white blood cells play a major role, we'll begin with them. And in order to do this we must look at bone marrow because that is where white blood involvement of phagocytes first. Phagocytes are cells that eat invaders, and they kill the invader in a five-step process. First, the phagocyte searches out the invader and responds to its chemical signals. In the second step the phagocyte attaches itself to the enemy. In the third step, called ingestion, the phagocyte extends projections called pseudo pods, which completely surround the pathogen. In the fourth step, digestion, the phagocyte releases chemicals, which begin destroying the enemy. In the final step enzymes and oxidants literally take the invader apart.

Now we will examine the tactics of natural killer cells. Instead of completely surrounding the enemy as do phagocytes, the NK cells actually shoot the enemy full of holes by releasing a chemical called perforin, which causes the pathogen to leak to death. The NK cells can also release molecules, which cause the interior of the target to shrink and finally break apart.

The innate system has a couple other defense mechanisms. Nasal hairs trap many foreign substances and tiny hairs in the respiratory system trap foreign substances heading for the lungs. We either cough or sneeze these substances out of our bodies. Also hydrochloric acid in the stomach kills many bacteria we swallow with our food.

Having looked at the innate defense system, we can now examine the adaptive immune system, which will include T and B cells and antigen-presenting cells. (T and B cells are known as lymphocytes.)

In the innate defense system the phagocytes indiscriminately attack invaders and destroy them as described earlier. In the adaptive immune system the lymphocytes must recognize the invader; they do not attack indiscriminately.

As noted earlier almost all blood cells originate in bone marrow but as lymphocytes develop they become either B cells or T cells. The earliest T cells multiply in the thymus, so they are called T cells. The B cells develop in the bone marrow so they are known as B cells. Each T and B cell develops chemical receptors on its surface, but each cell's receptors are designed to recognize only one enemy. The cell is not considered mature until it has used its receptors to attach itself to an antigen, its enemy. So each lymphocyte is programmed to search for its enemy, and millions of its "cousins" contain other chemical receptors and search for other enemies. In the bone marrow about one million lymphocytes are produced every second, so the human body always has plenty of B and T cells. Many lymphocytes will never see any action because their chemical receptors will never come in contact with the enemy they are programmed to attack. It is estimated that there are possibly 100,000 different chemical receptors searching for invaders.

The third kind of cell mentioned previously is the antigen-presenting cell (APC). These cells actually swallow microbes (disease causing bacteria) and other antigens, then signal T cells what they have eliminated. They have "presented" the antigen to the lymphocyte. Now the T cell causes the mobilization of more APCs to go after this particular group of antigens.

The lymphocytes circulate throughout the body increasing chances that they will encounter enemies. Some "establish residence" in lymph nodes in the lymphatic system and are ready to attack invaders as the lymph fluid is filtered in the lymph nodes.

There is something else we probably aren't aware of - cancer cells (figure 3) are in our bodies but we don't even know about their presence because many times our immune system can attack and destroy them without our ever knowing about it.

are taking place. Anytime our cells have attacked and destroyed an invader, our cells memorize the make up of the invader and are ready for its reappearance. For instance, before vaccinations were invented and dangerous diseases such as diphtheria invaded a body, if that body overcame the disease it developed its own immunity and the disease could not cause that body any more trouble.

Chapter 12, Figure 3

Attacking cancer cells
In this artist's sketch T lymphocyte cells (green) have identified a foreign cell (tan) in this case cancer, and destroy the invader. This could easily be the destruction of a cancer cell the person knew nothing about.

Chapter 12, Figure 4

Vaccine immunity
In this artist's sketch a white blood cell (tan) recognizes a foreign invader, in this case a tuberculosis vaccine (green). The vaccine causes the immune system to prepare for more of the same invaders so the body is immune to tuberculosis.

Vaccinations are a man-made means of protecting the human body against certain diseases. Scientists use dead or weakened cells of the disease, develop those cells into a vaccine, inject the patient with the vaccine and the body's disease destroying cells attack the invader (figure 4) and destroy it, memorize the disease and are ready to attack this disease when it shows up in the body again. And we don't even know that this invasion of the enemy microbe and its destruction

We now begin to understand how complicated the design of the immune system really is. We have not begun to cover all of the details but now we do know some of the basics. Obviously the medical student will dig much deeper.

Chapter 13
SMOKING

We have taken a brief look at how amazing and complicated the human body is. In spite of the fact that the design shows everything smoothly working together, we sometimes invite trouble because we don't think too clearly. One habit we fall into, probably at first because it makes us look/feel "cool", is the somewhat socially acceptable act of smoking.

Nicotine and carbon monoxide are just two of the harmful agents we ingest when we smoke. These are the main gases we inhale, but we also pull in over four thousand particles as we smoke, including tar. We see these particles, actually unburned tobacco, in cigarette smoke and they are known to poison cells, cause cancer, alter cell structure, suppress the immune system and alter neural activity in the brain. Tar puts a dangerous coating on the interior of our lungs and contributes to lung cancer.

The gases we don't see do their share of damage also. If we smoke we inhale carbon monoxide, which is also the poisonous gas in automobile exhaust fumes. This gas unites with blood causing less oxygen to be carried in the blood stream. This means less oxygen reaches all parts of the body, obviously causing cell damage. Nicotine works adversely on the central nervous system increasing blood pressure, constricting blood vessels and increasing heart rate. And nicotine is the substance that causes addiction - we all know many smokers who tell us they would like to quit but they just can't. And it probably started "innocently" enough as a teenager when he sneaked a cigarette and lit up because it was the "cool" thing to do.

The lungs are most affected by smoking. Tiny cilia, hair-like cells, that are in a constant waving motion in the air passages of the lungs trap particles that are eventually moved out of the air passages and into the throat where they no longer affect the lungs. But harmful substances in the smoke cause a kind of paralysis in the cilia making them slow down in their wave-like motion. The result is that these harmful substances in the smoke settle in the lungs and reduce their ability to send oxygen to the entire body. We have seen pictures of lungs belonging to smokers, lungs that are blackened over time by the harmful substances in cigarette smoke.

But the most dangerous effect smoking has on the lungs is cancer. The foreign substances in smoke eventually destroy the lungs to the point that they cannot perform as they are designed. These foreign substances finally can cause altered cell structure and cancer is the ultimate result. In the world's most developed countries smoking is the leading cause of death of people under age 65.

Emphysema is also a lung disease which is caused mostly by smoking, but polluted air, including dirty air from industrial sites, is also a reason a person may develop emphysema. The disease develops large air spaces in the lungs by destroying alveolar walls. Then the lungs cannot deflate properly to expel air during exhalation. As a result less oxygen is pulled into the blood stream, so people with emphysema become short of breath and when under any kind of physical stress they must breathe harder/more rapidly. Emphysema cannot be reversed but in some cases it can be stopped or slowed down by removing the cause, for instance, smoking.

In 1964 the U.S. Surgeon General's office issued a report officially confirming what scientists already knew - that smoking was a cause of lung and laryngeal cancer. Soon after that report the U.S. Congress passed legislation banning cigarette advertising on television and requiring health warnings on cigarette packages. Now many places prohibit smoking on their premises because of the damage of second-hand smoke.

There are many other ways we intentionally damage our bodies, but the U.S. government has taken definite steps to alert us to the damage caused from smoking.

Following are statements issued by the U.S. Surgeon General since 1960 concerning smoking:

- Cigarette smoking may be hazardous to your health. (1960-1970)

- The Surgeon General has determined that cigarette smoking is dangerous to your health. (1970-1985)

- Smoking causes lung cancer, heart disease, emphysema, and may complicate pregnancy. (1985 -)

- Quitting smoking now greatly reduces serious risks to your health. (1985 -)

- Smoking by pregnant women may result in fetal injury, premature birth and low birth weight. (1985 -)

- Cigarette smoke contains carbon monoxide. (1985 -)

In addition, the Surgeon General issued two more warnings - cigar smoking can cause cancer of the mouth and throat and is not a safe alternative to cigarettes; and smokeless tobacco can cause cancer of the mouth, gum disease, tooth loss and is not a safe alternative to cigarettes.

From a practical point of view, the U.S. Congress cannot legislate a ban on smoking. We would see all kinds of lawsuits going all the way to the Supreme Court. But our government has issued enough information/warnings that people can be expected to use some common sense and willpower to take good care of their bodies.

Chapter 14
CELLS, CHROMOSOMES, DNA AND GENES

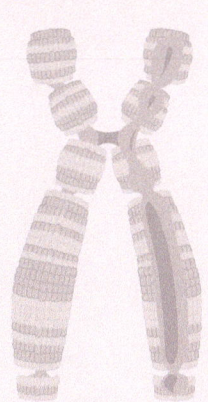

We've all heard the "explanation" - "His DNA matches the DNA found at the crime scene". That one statement convinces us that law enforcement has determined who the criminal is. And chances are overwhelmingly in favor of the accuracy of that conclusion. DNA doesn't lie.

So, what is DNA?

There are two definitions of DNA that are common among scientists. One is that "it is a library with the genetic information in a cell that allows that cell to exist, grow and reproduce itself." Another definition is that DNA is the "blueprint of life".

Defining genes is also important if we are to understand how genes and DNA are involved in the way our bodies work and survive, and even why we look the way we do. Genes are part of a DNA molecule and simply put, genes determine our sex and all of our physical characteristics. We inherited much of what our parents are because we inherited their genes.

But other things are also at work. Cells (figure 1) are the microscopic, living things that make up all of the different parts of our bodies. Cells that make up a specific body part are different from the cells of all other body parts. Chromosomes (figure 2) are also enclosed in the nucleus of cells and they are composed of DNA molecules. In the human body there are 46 chromosomes in each cell. Red blood cells have no chromosomes.

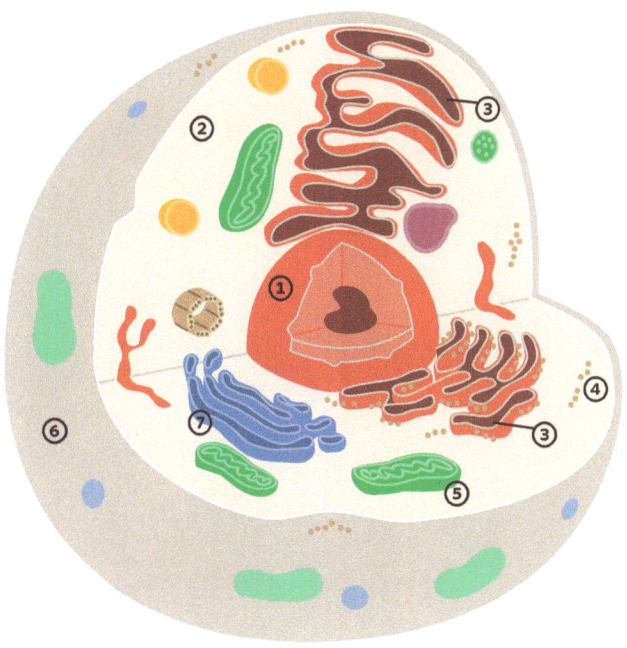

Chapter 14, Figure 1

The human cell
This drawing of the human cell is only intended to give us an idea how detailed and complicated it is. The cell is a microscopic part of the body. The nucleus is the center part of the cell and is much smaller than the entire cell. In the nucleus are forty-six chromosomes arranged in pairs of two. These chromosomes are composed of DNA strands wrapped around a protein core. Genes are in this composition and Scientists estimate there at least twenty thousand genes in the human body.
1. Nucleus – Chromosomes containing DNA and genes are found in the nucleus.
2. Cytosol - Fluid contained in the cell.

3. Endoplasmic reticulum - Molecules are transported and stored here. This endoplasmic reticulum is a collection of tubes, membranes and sacs which either store or transport molecules.
4. Ribosomes - Responsible for the production of proteins.
5. Mitochrondria - Energy is produced in these tiny parts of the cell. Sugars and fats along with oxygen are broken down to provide energy used by the cell. Respiration in the cell happens here.
6. Plasma membrane - covers the cell. Any materials moving in or out of the cell must pass through this membrane.

Chapter 14, Figure 2

Chromosomes and Genes
Chromosomes are composed of strands of DNA. There are forty-six chromosomes arranged in pairs of two in each cell. Children inherit one of each paired chromosome from each parent. So a child gets forty-six chromosomes, but twenty-three came from each parent. And in those chromosomes are the genes, which determine which characteristics the child inherits from each parent.

So in a cell chromosomes hold DNA and genes are actually strands of DNA and this "mixture" is passed from parents to children. The chromosomes in each cell are arranged in 23 pairs. Only one chromosome of each pair is inherited from each parent. So a child inherits 46 chromosomes but half come from each parent. Because of this a boy doesn't look exactly like his father and a girl doesn't look exactly like her mother. (The drawing of the cell in figure one emphasizes the complexity of this microscopic part of the human body.)

But one of the most interesting points about all of this is that the DNA of every human being is different from that of every other person. (In the case of identical twins their DNA will be the same.) In the science lab a scientist can identify the DNA of any person, knowing that he/she is looking at a pattern that can identify only one person. No other person will have that same DNA pattern.

DNA can also identify who a person's parents are because certain characteristics of DNA patterns follow a line of descendants.

Genes are defined by realizing that they are sections of DNA strands. The protein core in figure two is not part of a gene. That core simply gives the DNA a solid foundation. To get an idea about the makeup of the interior of the cell's nucleus consider this. There are 46 chromosomes in a cell's nucleus. Each cell holds all of the genes in the human body, and scientists estimate that there are at least 20,000 genes in our bodies. By dividing the number of chromosomes into the number of genes we calculate that there are more than 400 genes in each chromosome! Realizing that cells are microscopic and that chromosomes are just tiny parts of cells and there are forty six of them in the nucleus, we begin to realize how unimaginably complex our bodies are. And as mentioned earlier - the cells in any part of the body are different from the cells of all other body parts.

Something else worth noting - red blood cells and platelets have no chromosomes and male and female reproductive cells have only 23 chromosomes - just one from each pair.

The science and chemistry of understanding DNA, genes and chromosomes is beyond the scope of this book, but it is interesting to see a drawing of a computer generated image of DNA in what is called the classic helix arrangement (figure 3).

One conclusion is worth considering - the human cell is a complicated, microscopic piece of our bodies and its composition makes us aware of complex design. And we wonder - is there intelligence behind this design?

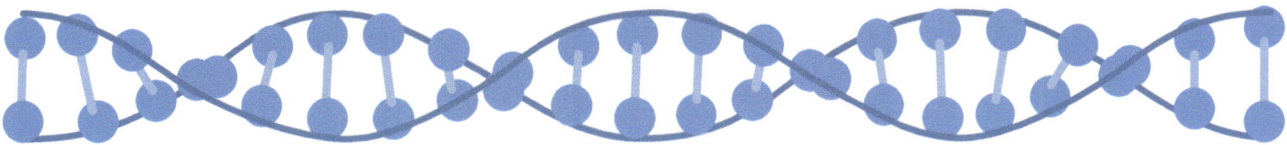

Chapter 14, Figure 3

DNA

This is a drawing of a computer-generated image of DNA in the classic helix arrangement. This is only one example of how DNA would appear in the human cell under an electron microscope.

Chapter 15

WHAT MOTIVATES THE MEDICAL STUDENT?

If you have read through most of this book you have probably wondered what the life of the medical student is like, what motivates him/her and maybe how late in a college career a student can decide to enter the field of medicine. Although this book was not meant to be part of the medical student's curriculum/course of study, there will be many questions raised about the field of medicine. It becomes quite apparent that a doctor must be able to absorb just about all of the information available about the human body.

We will take a brief look at pre-med courses but first we can identify the student who probably will be motivated to become a doctor. Two traits stand out in all who enter this field - the desire to heal and the desire to serve. Interest in science is also an aptitude that helps but the desire to learn how to heal and serve are way out front. We also find out that almost all doctors are self-starters and they know they must always be willing to learn. A successful, effective doctor never quits learning. A pre-med student can actually be a liberal arts major.

There are four courses a college student must complete if he/she intends to one day be a doctor - a year of organic chemistry (dealing with the world of molecules), one year of physics, a semester of calculus and a year of general chemistry. These are minimum requirements. Excellent grades and having shown an interest in the medical field show that motivation has moved in the right direction.

The serious pre-med student is also thinking about medical school and knows pretty well that competition to get into the "best" schools will be rather fierce. Medical schools like to know that potential students have excelled in something that required real effort, and volunteer work shows a willingness to serve.

MEDICAL SCHOOL

Medical schools require that a potential student generate excellent recommendations from people who know the student well. There is also the MCAT - the Medical College Aptitude Test. The recommendations, the score on MCAT and grades are all important. In addition there is the all-important interview. Being accepted into medical school obviously is tougher than getting into college as a freshman right out of high school. There are at least a couple reasons that cause potential students to be rejected - poor grades and poor interviews.

Once accepted the student is headed for seven years of intense study but the highly motivated students keep their focus on that degree at the end.

The first year of study gives the student a good sense of the make-up of the human body because a cadaver will receive the student's full attention for a full year, working as a team with other students. They will actually dissect the body and when the year is up they will have memorized at least eight thousand terms most of which they will probably use the rest of their careers. Class work will involve microbiology, pharmacology and immunology.

The second year requires more class work, but the student will also have contact with patients and will observe autopsies required when a person has died from unknown reasons. Pathology is also a class the student will take.

The third year the student begins rotating through specialties such as cardiology, urology and other specialty areas. About six weeks are spent in each of these specialty areas.

In the fourth year the student begins to direct himself/herself into the area which will become a lifelong career. Also during this year the student begins selecting where he/she wants to become a resident physician, but still a student.

In the fifth, sixth and seventh years, the medical student selects the field he/she will work in - specialty or internal medicine or family medicine. Family medicine includes all age groups while internal medicine generally serves those age 16 and older. The internship is served the fifth year and the residency covers the sixth and seventh years.

The medical student has spent four years in college, which is considered pre-med, four years in medical school and three years as an intern and resident physician, still in training and still being supervised. But at the end of this seventh year of medical school he/she becomes a physician with a degree and ready to practice medicine full-time.

Trying to understand the motivation that keeps medical students going is not easy. They spend much time studying and working. They endure many hours without enough sleep because of the pressure and rigors of the courses of study. By the time they have finished four years of pre-med and another four years of medical school they know they face three years as interns. They have an idea what lies ahead, but they are still in for some surprises. However, after eleven years of studying, working and even dreaming they finally attach the title "physician" to their name.

As you have looked at many of the drawings in this book, you can realize no matter how complicated they appear, your doctor can identify every part and can tell you how everything fits together and how everything works.

So when we talk with the doctor in his/her office or spend time in the examining room or face surgery and maybe even face some difficult decisions of our own, we can rest assured that someone who has really been "through it" is now perfectly qualified to help us and for this we should be thankful to say the least.

SUMMARY STATEMENTS

The following summaries encourage us to consider design versus chance.

We now have a pretty good grasp of how the human body is put together and how it functions. The issue is simply this - did chance and uncontrolled accidents make the body what it is or did a source of intelligence design this amazing "machine"?

We don't need to review all that we have just covered, but a few concise summary statements can help us answer the question posed in the title of this book, "Intelligent Design or Non-intelligent Design"?

So keep the book's title in mind as you consider the following summary statements:

- In the top part of the heart's right atrium is a cluster of cells, the Sino-atrial node - also known as the pacemaker. This pacemaker sends signals to each atrium telling each when to beat. It also sends signals to the atrioventricular node on the floor of the right atrium, which in turn tells the ventricles to beat. The ventricles beat about one tenth of a second behind each atrium. Figure four in chapter one shows these two nodes in the heart.

- One of the body's senses, hearing, required perfectly placed openings through each side of the skull so that the outer ear, middle ear and inner ear could follow a path designed for the transmission of sound waves that become electrical impulses all the way to the brain. Either these openings were designed into the bone or they are there by chance. And the inner ear sits inside a protective bone "cavern" giving additional protection.

- In the middle ear are three tiny bones that connect the outer ear to the inner ear. These bones send vibrations to the inner ear and finally to the brain. See figure one in chapter seven. When sound waves move through the inner ear, they are changed into electrical impulses before they move on to the brain. The part of the brain that interprets sound needed electrical impulses instead of waves of sound, so the inner ear does what it is designed to do.

- The brain sends signals to the diaphragm and rib muscles telling them to create space for the lungs to expand, allowing the lungs to inhale. Then, after the lungs are full of air, the brain sends a signal to the rib muscles and diaphragm to relax. This relaxing compresses the lungs causing them to exhale. The lungs have no muscles of their own. Without those signals from the brain the lungs could not function. And even when we are asleep the cooperation between the brain and the diaphragm & rib muscles keeps us alive although we are completely unaware of the action.

- Blood in the veins in the legs is affected by gravity more than in any other part of the body. And gravity works against that blood as it is pumped toward the heart, and would pull it back toward the feet. But one-way valves in the veins close down, blocking the backward flow of blood so the blood can only move toward the heart. Those one-way valves are placed in the veins as an absolute necessity.

- The skull is the layer of bone that protects the brain. But between the skull and brain are three layers of protective tissue, meninges, one filled with fluid that further protects the brain from damage from outside blows. These same kinds of protective tissue line the spinal cord, giving it the same protection as the brain. The spinal cord conducts messages to and from the brain and must have the protection provided by the meninges and spinal column.

- The bones in the spine are called vertebra. Each has an opening through which the spinal cord extends from the lower part of the brain to its final destination. These openings are basically the same in all vertebra, and the spinal cord had to have the protection it receives inside the spinal column. Otherwise it would have been a freestanding nerve package subject to all kinds of abuse.

- At the back of the mouth the trachea (wind pipe) takes air to the lungs, and the esophagus takes food to the stomach. When a person swallows food a structure called the epiglottis covers the wind pipe so that food can only move through the esophagus to the stomach. The epiglottis is designed to prevent choking, and it only closes when swallowing takes place.

- In the lungs blood carries carbon dioxide to alveolar sacks - about one million per lung, and oxygen is exchanged for the carbon dioxide which is exhaled. See figure three in chapter four. Capillaries on each alveolar sack manage the exchange of carbon dioxide and oxygen, and the blood is programmed to give up carbon dioxide and receive oxygen. The body cannot live without this exchange in the lungs.

- The internal structure of the kidneys shows there are about one million nephrons per kidney. The nephrons filter waste products from the blood and finally send "cleaned" blood back to the circulatory system. The nephrons also begin the task of sending the remaining "dirty fluid" - urine - to the bladder. The nephrons are designed to perform an obviously important function and their removal of waste must happen to help keep the body alive.

- The diaphragm is a sheet of muscle that separates the abdominal and thoracic cavities (the upper and lower parts of the body). As mentioned, contraction of the diaphragm downward helps cause the lungs to fill with air, and when it relaxes it helps the lungs expel air. But the diaphragm also has three openings - one for the esophagus, one for the main artery and one for the main vein. These openings are designed for a reason, they did not happen on their own.

- The gall bladder stores bile, which is manufactured by the liver. Bile is necessary in the breakdown of fats. When the stomach senses that fats are in the food it holds, it sends a signal to the gall bladder to release bile into the duodenum, the top part of the small intestines. Releasing bile is just one of many processes that take place in the entire digestive system. The liver makes the bile from dead red blood cells.

- Looking at the brain from the top, there are two halves which make up the biggest part of the brain, the cerebrum, shown in figure two, chapter three. When the brain receives a signal from any part of the body it processes that signal immediately and sends a response to the area the signal came from. This happens in microseconds. The central nervous system, including the brain, was covered in chapter three. The central nervous system is an ingenious assemblage of millions of nerves to make it the most meticulously designed control center on this planet.

- The liver receives from the small intestines all of the nutrients from the food that has been digested. But the liver has more to do than just work on those nutrients. In fact, the liver completes at least five hundred different tasks that help keep the body healthy and functioning. The liver continues to function even while the body is at rest/asleep.

- At the back of each eyeball are optic nerves. The optic nerves taking signals from the left half of each eye take those signals to the right part of the brain. The optic nerves taking signals from the right half of each eye take those signals to the left part of the brain. The point at which two of the four nerves cross each other is called the optic chiasma. When those signals reach the back part of the brain they are interpreted - it is only then that we know what we have seen and the brain actually turns an inverted image right side up. Figure three in chapter nine shows the optic nerve.

- Capillaries are tiny connectors between arteries and veins. Capillaries "allow" oxygen to be removed from arteries and they "allow" carbon dioxide to enter the veins. Their design is important and the body could not live without these microscopic "gas exchangers".

- The roof of the mouth is called the palate. The hard palate is in the front and the soft palate is in the back part. When we have chewed our food and begin to swallow, the tongue pushes the food up against the soft palate closing off the nasal passages so that the food moves down the esophagus and not up into our nasal passages. That design helps us swallow our food.

- The aorta is the artery that takes blood from the heart and "sends" it to all parts of the body. We think of the heart as the muscle that keeps us alive. But as the aorta leaves the heart three smaller arteries branch off and actually send blood to all of the surfaces of the heart, front and back. The heart also must have oxygen-rich blood for its own existence. The heart's construction shows that this blood supply system had to be in place and functioning as soon as the heart started beating. When we hear that someone has experienced by-pass surgery, it is these surface arteries that were involved.

- In the body's immune system, covered in chapter 12, there are cells which attack "foreign invaders" and destroy them by either eating them or shooting them full of holes. There are also cells that destroy the invader and signal other similar cells to be on the lookout for these enemies. Some of the disease-fighting cells are programmed to attack any foreign invader; others are programmed only to attack one specific kind of an enemy. Then there are also the tiny hairs in the trachea which are constantly waving toward the opening of the trachea and moving foreign objects to the top of the trachea to be coughed up to the mouth and then out of the body.

- The lymphatic system is a one-way conducting system, separate from arteries and veins. Capillaries leak a slight amount of lymph fluid, which is absorbed by the lymph vessels and is conducted to lymph nodes where it is filtered to remove harmful substances, then the filtered fluid returns to the blood stream. Chapter twelve, figures one and two show the conducting system and a typical lymph node.

- As we examine the art accompanying chapter 14, we realize how complicated the design of the body is and that this design is necessary so that the body can function as it must. And we can ask ourselves if all of this happened by chance or accident, or is there some kind of planning behind it?

- In the nucleus of the cells mentioned above there are forty-six chromosomes and these chromosomes contain DNA and genes. The chromosomes are arranged in pairs of twenty-three. However, male and female reproductive cells have only a total of twenty-three chromosomes - no pairs. But when the male and female cells unite, the new embryo's cells now have forty-six chromosomes, just like its parents.

This process shows both planning and design, but some will try to convince us that it all happened by chance or accident, and even if it does show design, there was no intelligence behind the design. And that gives us the term "non-intelligent design".

These summaries could cover much more than what you have just reviewed, but we do begin to understand why it makes sense to say the human body is designed and did not put itself together by accident and/or chance. Common sense tells us that when a need is identified, an intelligence greater than our body parts had to recognize that need and construct whatever was necessary to satisfy that need. It simply does not make sense to claim that the human body needed its hundreds of parts and needed them all at the same time, so the body with no outside help, designed those parts, put them in place and immediately got them working as they must. Anytime a body part shows design there has to be intelligence behind that design.

Going back to that "primordial soup", we are told to accept, as fact, the theory that non-living objects banged together and became living cells. We are also told to believe that, although design is overwhelmingly apparent, there is no intelligence behind that design. We are told to believe that all living things that we see, including all human beings, are the results of chance and accident. We also are probably encouraged to believe that we humans are the most intelligent beings in this vast universe. Does common sense possibly tell us otherwise?

Readers will notice that this book does not in any way attempt to control or manipulate a reader's thinking about religion. People can feel that both evolution and intelligent design are theories that need to be investigated. Then the individual can decide which, if either, theory to accept.

ACKNOWLEDGMENT

Dr. David Edwards, with Sterling Primary Care at Centennial Hospital in Nashville, Tennessee, served as consultant to Ericson throughout the writing of this book. Dr. Edwards has read the entire book and has suggested changes/corrections as necessary. He recognizes that evolution is theory instead of fact, as most doctors do, and became the de facto "Consulting Physician". His discussions with the author about the medical profession became the basis for chapter fifteen about the men and women who go through the rigorous training to become physicians.

TEST ON INTELLIGENT DESIGN

Take this test before you read this book, then again after reading it. See what you have learned. You get three points for each correct answer and one point for taking the test.

Suggestion: On a sheet of paper, place numbers 1-33 and put your answers on that sheet - save the test so that other family members/friends can also take it without being overly influenced by your answers. (You might even be wrong!) The answers follow the test. There is only one correct answer for each question unless all of the answers are correct, then mark (d) only.

1. The sinoatrial node is (a - the largest taste bud on the tongue; (b - the hind-part of the brain; (c - the scientific name for the bone in the ball of the foot; (d - part of the heart)

2. Arteries carry (a - lymph fluid to the liver; (b - blood from the heart; (c - articulated fluid from bone joints; (d - none of the above)

3. White blood cells (a - make up over fifty percent of the blood; (b - fight disease; (c - transport oxygen to the lungs; (d - form the inner lining of the eyeball)

4. The optic chiasma (a - focuses light on the retina; (b - is another name for the lens; (c - is the fluid in the eyeball; (d - is the point at which optic nerves cross each other)

5. The term pia mater relates to the (a - diaphragm; (b - kidneys; (c - tongue; (d - brain)

6. The middle ear contains (a - three bones related to balance; (b - a snail-like structure that sends electrical signals to the brain; (c - special nerves that send information to the brain; (d - none of the above)

7. Blood flows from the right side of the heart to (a - the aorta; (b - the pancreas; (c - the lungs; (d - the spinal cord)

8. Nephrons are in the (a - blood stream; (b - lungs; (c - spinal cord; (d - kidneys)

9. In the eye the lens is (a - shaped by muscles; (b - attached directly to the optic nerve; (c - influenced by the spinal cord; (d - all of the above)

10. Rods and cones are found in the (a - ear; (b - eyes; (c - throat; (d - esophagus)

11. Alveolar sacs are found in (a - the lungs; (b - small intestines; (c - the heart; (d - the tongue)

12. Food is digested mostly in (a - the stomach; (b - the small intestines; (c - large intestines; (d - the blood stream)

13. The inner portion of bones is referred to as (a - matrix fluid; (b - marrow; (c - meninges; (d - none of the above)

14. In order for our sense of vision to perform as it is designed, our eyes depend on part of the brain to interpret what is in our field of vision. That part of the brain is located in the (a - front; (b - back; (c - top; (d - bottom) of the brain.

15. Concerning our sense of vision, the part of the brain that interprets what we have seen also (a - controls blood pressure; (b - turns an inverted image right side up; (c - interprets sound waves; (d - responds to our sense of smell)

16. Phagocytes are (a - a special type of taste bud; (b - part of the immune system; (c - the outer covering of our teeth; (d - chemicals in the roots of our teeth)

17. The eyeball (a - receives blood through an opening in the back; (b - is filled with fluid; (c - is about one inch in diameter; (d - all of the above)

18. Arachnoid mater (a - helps protect the spinal cord; (b - protects the eye from excessive light; (c - contains fluid that attacks venom from spider bites; (d - is not found in the human body)

19. The middle ear (a - contains organs of balance; (b - sends nerve signals to the outer ear; (c - contains the sinoatrial node; (d - contains three tiny bones)

20. Lungs expand because of (a - internal muscles; (b - contraction of the ribs; (c - meninges; (d - action of the diaphragm and ribs)

21. The trachea is (a - the lower part of the esophagus; (b - a bile duct; (c - the wind pipe; (d - a protective lining of the brain)

22. The epiglottis (a - controls the amount of air that can enter the nasal passages; (b - is part of the cerebellum; (c - keeps a person from choking; (d - none of the above)

23. Alveolar sacks (a - provide for the exchange of carbon dioxide and oxygen; (b - are involved in the production of urine; (c - are located in the bladder; (d - help protect the spinal cord)

24. From the lungs, blood (a - goes directly to the kidneys; (b - is deoxygenated; (c - goes to the heart; (d - has been processed by nephrons)

25. Platelets are found (a - on the surface of teeth; (b - in the blood stream; (c - supporting the optic nerve; (d - in T cells)

26. The liver (a - prevents the bladder from manufacturing urine; (b - is located above the diaphragm; (c - turns certain blood cells into bile; (d - receives food from the esophagus)

27. The stomach (a - is a muscular sack; (b - is the first organ that begins digesting fats; (c - has a lining that is replaced about every three days; (d - all of the above)

28. In the brain (a - the pons; (b - the cerebellum; (c - the medulla oblongata; (d - the cerebrum;) is divided into two halves)

29. The lungs (a - are attached directly to the ribs; (b - can breathe on their own because of internal muscles; (c - are not shaped the same because of the heart; (d - all of the above)

30. Between the skull and the brain, there are (a - three; (b - four; (c - five; (d - none of the above;) protective linings.

31. If it were not for capillaries (a - carbon dioxide would not flow to the heart; (b - carbon dioxide could not reach the lungs (c - oxygen could not be made available to the body's cells; (d - all of the above)

32. The terms gyri and sulci are associated with the (a - optic nerve; (b - large intestines; (c - tongue; (d cerebrum)

33. When a person swallows (a - the esophagus restricts the amount of food that can enter the stomach; (b - the sense of taste cannot function until the food has reached the stomach; (c - the epiglottis and soft palate perform protective tasks; (d - all of the above)

ANSWERS TO TEST

1. (D) chapter 1
2. (B) chapter 1
3. (B) chapters 2 and 12
4. (D) chapter 9
5. (D) chapter 3
6. (D) chapter 7
7. (C) chapters 1, 2 and 4
8. (D) chapter 6
9. (A) chapter 9
10. (B) chapter 9
11. (A) chapter 4
12. (B) chapter 5
13. (B) chapter 10
14. (B) chapter 9
15. (B) chapter 9
16. (B) chapter 12
17. (D) chapter 9
18. (A) chapter 3
19. (D) chapter 7
20. (D) chapter 4
21. (C) chapter 4
22. (C) chapter 5
23. (A) chapter 4
24. (C) chapter 4
25. (B) chapters 2 and 12
26. (C) chapter 5
27. (D) chapter 5
28. (D) chapter 3
29. (C) chapter 4
30. (A) chapter 3
31. (D) chapters 2 and 4
32. (D) chapter 3
33. (C) chapters 5 and 8

www.ingramcontent.com/pod-product-compliance
Lightning Source LLC
Chambersburg PA
CBHW050746180526
45159CB00003B/1363